CW01654065

184
£5.99
0509

John Wesley:
travelling geographer

John Wesley

John Wesley: travelling geographer

Howard Oliver
Former Lecturer and Supernumerary Fellow
Harris Manchester College
Oxford

With an introductory chapter by
Ralph Waller
Principal
Harris Manchester College
Oxford

2015
Harris Manchester College
Oxford

Published by
Harris Manchester College,
Mansfield Road,
Oxford OX1 3TD

First published 2015

© 2015 Howard Oliver
© 2015 Ralph Waller (Chapter 1)

The right of Howard Oliver and Ralph Waller to be identified as authors of this work has been asserted by them in accordance with sections 77 and 78 of the Copyright, Designs and Patents Act of 1988.

All rights reserved. No part of this book may be reprinted or reproduced or utilised in any form or by any electronic, mechanical, or other means, now known or hereafter invented, including photocopying and recording, or in any information storage or retrieval system, without permission in writing from the publisher.

ISBN 978-0-9555098-5-8

A catalogue record for this book is available from the British Library

Typeset in Baskerville
by Deltatype Ltd, Birkenhead, Merseyside

Printed in Great Britain by
Berforts Information Press Ltd, Eynsham, Oxford

Dedicated to
the memory of Howard Oliver's mother
Lucy (1920-2014)
A regular Methodist Chapel goer

Contents

Introduction

John Wesley lived through most of the eighteenth century and was among the most widely travelled people of his day, making innumerable journeys through Britain and Ireland as well as spending a time in America and visiting the European Continent.

He was a prolific writer both of religious works (sermons, hymns, addresses, periodicals), but also wrote on a wide range of other topics. His primary aim was to provide affordable religious, educational and informative books for his congregations. He also published a comprehensive *Journal* describing his travels as well as writing a continuous stream of letters to both senior figures and ordinary people.

This book presents a selection of his writings, mainly from the *Journal* of his travels and some letters, covering a whole spectrum of geographical topics, both physical (scientific), human and descriptive. The aim is to provide the reader with a vivid picture, not only of the travels of Wesley himself, but also of the landscape, life, times and environment in the eighteenth century as he observed them.

The authors hope that reading this book will help to bring the real person of John Wesley to life and thus help in appreciating any other study of his life and teachings, and provide a useful resource for articles and sermons on Wesley. It may also stimulate reliving some of the events by visiting places he so vividly described.

Dr Howard R. Oliver
Former Lecturer and Supernumerary Fellow,
Harris Manchester College,
Oxford

Revd Dr Ralph Waller
Principal,
Harris Manchester College,
Oxford

I

A portrait of the man and his times

John Wesley was one of the cleverest people of his age, who engaged with many of the great issues of his day. Like other eighteenth-century intellectuals, he read widely and thought deeply, but he also wrote and preached on some of the major problems facing his generation. But unlike many of his contemporaries, who stayed within their own social strata, he was acquainted with a great number of men and women from a wide variety of backgrounds: intellectuals such as Dr Johnson, the leading man of letters of his day, and Bishop Butler, the eighteenth-century philosopher. He corresponded with politicians such as Pitt and Wilberforce, and was a regular visitor at the home of Selina, Countess of Huntingdon. He regularly visited the prisoners in Oxford, Newgate, and Marshalsea prisons, and his preaching touched the lives of the harlots of Drury Lane, the tinners of Cornwall, the keelmen of Newcastle, the smugglers of Devon, and the miners of County Durham.

Wesley, perhaps more than any other person in the eighteenth century, had his finger on the pulse of the nation. He travelled extensively through the British Isles, covering an average of 6,000 miles a year; talking with those he met, staying in the homes of his acquaintances, and bringing news and medical remedies with him. His *Journal* chronicles the countryside through which he passed and the weather he encountered. His writings provide us with a unique snapshot of life in the British Isles and are a source of interest, even for those who are not concerned with his religious beliefs.

John Wesley was born at Epworth in Lincolnshire in 1703, the seventeenth child of his parents. His father Samuel, the Rector of Epworth, had been brought up as a Non-Conformist, but during his studies at Newington Green Dissenting Academy he was asked by the Principal, Charles Morton, to write a paper refuting the doctrine of the Church of England.[1] Samuel's wide reading and thoughtful preparation for this essay led him to the conclusion that it was Nonconformity which had erred, and that the doctrine of the Church of England represented the true faith. He promptly turned his back on Dissent, accepted the Thirty-Nine Articles, and enrolled as an undergraduate at Exeter College, Oxford with the hope of becoming a minister in the Church of England. Similarly, John's mother, Susanna, had also been brought up as a Nonconformist. Her father, Dr Samuel Annesley, like many Anglican ministers, had been ejected from the Church of England in 1662 for his opposition to the compulsory use of the *Book of Common Prayer* at church services. From that time onwards until his death in 1696 he had presided over London Nonconformity. Susanna, his daughter, had grown up in the family house in Spital Yard, listening to the discussions around her father's dining table on the great issues of the day; free will and determinism, Calvinism and Arminianism. A frequent visitor to the house was Richard Baxter, who strongly advocated a catholic (meaning inclusive) approach to religion, in opposition to those who wanted to make Christianity more exclusive. Susanna, like Samuel, her future husband, weighed the merits of Nonconformity against those of the Church of England, and quietly abandoned the former in favour of the latter.[2]

Although John's parents agreed on the centrality of the Church of England in the religious life of the nation, they were both very forceful people who held strong and independent opinions on many issues with the consequence that their home was often a place of argument and tension. One evening while the Rector was saying prayers, he prayed for King William but, to his astonishment and

[1] Clarke, Adam, *Memoirs of the Wesley Family*, 2 vols. (London, 1836), Vol. 1:97

[2] Charles Wallace Jr, ed., *Susanna Wesley: The Complete Writings*, (Oxford, New York 1997), p.71

dismay, his wife refused to say 'amen' for she firmly believed that the true King of England was King James II and his heirs. The Rector flew into a rage and announced, 'Two kings, two beds', and promptly left for London.[3] While he was away, Susanna robustly conducted services of worship in the Rectory kitchen. These services proved to be much more popular than her husband's services in the Church.

However, when the Rectory caught fire in 1702, he immediately returned home from London to be united with his family in their time of need. The child of Samuel's and Susanna's reconciliation was John Wesley.

As well as arguments between two strong-minded parents, family life for the Wesleys at Epworth was also overshadowed by debt. This was caused by Samuel's total inability to manage money, together with the increased expenses of providing for the needs of a growing family. The portrait of Samuel spending his money before he had earned it, and of Susanna valiantly trying to make ends meet, does not altogether reflect the true picture. Susanna herself acknowledges that she also must take some responsibility for the poor family finances as she spent too much money on her dresses. Although Samuel made gallant efforts to increase his income through his writings and his search for patronage, and by taking on the additional parish of Wroot, life for him became a continual financial struggle, made worse by a second fire which destroyed the Rectory in 1709.

However, this was not the whole story of life in the Rectory for the Wesley children; they also enjoyed a first class education provided by their mother, access to Samuel's extensive library, his continual encouragement, and the enjoyment and support of one another's company. Letters written between the children in later life, while not always showing agreement, generally reveal a deep affection for each other. This was especially true of John and Charles, who

[3] Clarke, *Memoirs of the Wesley Family*, Vol.1:198

not only exerted a unique leadership of two brothers in the history of the Christian Church, but, in spite of their several differences over ordaining men for the American ministry and Methodism's relationship with the Church of England, they showed a genuine care, concern and affection for each other.

Charterhouse

At the age of eleven, John Wesley was offered a place at Charterhouse through the sponsorship of a family friend, the Duke of Buckingham. The boys of Charterhouse seem to have experienced far less bullying and brutality than was generally the case for pupils in public schools before the reforms enacted by Arnold of Rugby and Thring of Uppingham. This was due in part to the fact that Charterhouse was founded by Thomas Sutton in 1611 for the education of poor boys. Every pupil had a personal sponsor, and each child's fees were covered by a scholarship. As a small child in the school Wesley had some of his food taken from him by the older boys. However, when he looked back on his school days some fifty-seven years later, he came to the conclusion that the other boys had done him a good turn, as eating little was a good recipe for a long and healthy life.[4] This, combined with his father's advice to run round the school grounds several times a day, also served him well and gave him a lasting interest in taking regular exercise, which he believed contributed to his good health and long life.

There is considerable evidence that Wesley worked hard at his studies and was praised by his older brother, Samuel, for the progress he was making in Hebrew. He must have held a great affection for Charterhouse for, in later life, he would go out of his way to visit the school once a year, and in 1727 acted as a steward on Founder's Day.

[4] John Wesley, *Journal*, 28 June 1770

Christ Church, Oxford

Wesley became a student at Christ Church, Oxford in 1720. We know little about this period of his life, but what we do know comes mainly from four surviving letters written to his mother, and two of her replies.

It appears that John pursued his studies vigorously under the tutorship of the Reverend George Wigan, who was widely known as a 'great and good tutor' with an expertise in English language and Hebrew. Following Wigan's retirement to concentrate on his responsibilities as Rector of the parish of Old Swinford in Worcestershire, the Reverend Henry Sherman took over as Wesley's tutor. In the great tradition of Oxford tutors, Sherman not only took responsibility for Wesley's studies but also was a good friend and mentor to him, occasionally joining him for breakfast and giving Wesley a large number of books that he felt would be helpful to the younger man.[5]

Wesley's finances were always precarious during his early years at Oxford. From his letters, it appears that as well as the occasional gifts from his father, and an annual grant from Charterhouse of £10 per year, he was also helped by the generous support of his elder brother, Samuel, who was a master at Westminster School. It is interesting to note that, in his Christ Church days, and indeed for the rest of his life, he had long hair in order to save money on haircuts.[6]

Lincoln College, Oxford

In 1724, at the age of 21, John Wesley graduated from Oxford with a Bachelor of Arts degree, but stayed on in his rooms at Christ Church in order to study for a Master of Arts degree.

[5] Tyerman, L., *The Life and Times of the Reverend John Wesley*, 3 Vols. (London, 1890) Vol.1:50

[6] Green V.H.H., *Young Mr Wesley*, (London, 1963) p.67

Wesley was ordained as a minister of the Church of England in 1725 and, in the same year, the Wesley family learned that there would be a vacant Fellowship at Lincoln College, Oxford; this was one of the three Fellowships reserved, in the first instance, for natives of Lincolnshire. Samuel Wesley, John's father, who had always been an admirer of the academic life, threw himself wholeheartedly into raising support for John's candidature; he visited Dr Morley, the Rector of Lincoln College, who also held the living of the Lincolnshire village of Scotton, he canvassed the Bishop of Lincoln who was the College Visitor, and Sir John Thorald, whose son was vacating the Fellowship. John Wesley was duly elected to the Fellowship on 17 March 1726 and a jubilant Samuel wrote: 'Whatever I am, my Jack is a Fellow of Lincoln.'

One of the happiest moments in taking up a new post is the period between accepting the appointment and starting the duties. This period for John Wesley extended from March to September 1726. At Easter that year he set out to walk from Oxford to Epworth. He had always enjoyed walking and in old age attributed his robust health to this activity amongst others. For him it was the beginning of a long, happy and idyllic summer. He enjoyed the company and companionship of his sisters. He worked in the garden, he visited friends in the neighbourhood of Loughton, Bawtry, Doncaster, Gainsborough and Owston Ferry, and preached in local churches. He swam in the river, picked fruit, gathered in the hay with his sister, and collected flowers. For John Wesley, it was one of the happiest times of his life, which became even more memorable because of his close friendship with Kitty Hargreaves. They spent much time together that summer, and she is frequently mentioned in his diary. They went on walks, held hands and she often visited the Rectory, in fact so often that she was on one occasion chased away by Samuel, who would have been only too aware that if John married he would have to relinquish his newly acquired Fellowship at Lincoln College.

The long summer came to an end in September when John

returned to Oxford to take up his Fellowship at Lincoln College. The Fellowship provided John with the security of a stipend, accommodation and food, a freedom to follow his own interests, a standing in the community, and the opportunity to read widely, broadening his outlook and formulating his own ideas. Of the hundreds of books he read during the early years of his Fellowship, three in particular made a great impression upon him: Jeremy Taylor's *Holy Living and Holy Dying* reinforced his own belief in the forgiveness of sins through Christ, as a state of change conferred on the individual. This idea was important in his later work amongst prisoners and thieves, and harlots, in that he came to realise that no-one was beyond redemption, no matter what they had done, and that forgiveness was open to all. The second idea set out in *Holy Living and Holy Dying* which influenced him was Taylor's view of the importance of the purity of intention. This caused Wesley to resolve to dedicate all his life to God; as well as his thoughts, words, and actions.[7] Thirdly, Taylor influenced Wesley's ideas on the use of money in that he clearly sets out that whatever money individuals have which is not required by them should be given to the poor, a view which Wesley strongly echoed in several of his sermons.

The second book which influenced him was Thomas à Kempis's *The Imitation of Christ* which confronted Wesley with the nature of inward religion, and that outward works were not sufficient in themselves.[8]

The book which had the greatest impact on Wesley was William Law's *Serious Call to a Devout and Holy Life*. Wesley's diary and *Journal* reveal that he read the book many times, not only to himself but also aloud to those in his pastoral care. Many of the ideas contained in this book became central to Wesley's thought, especially the ideas propounded by Law on the importance of singing as a way of developing the religious life, which became a key strategy for

[7] Wesley, John, *A Plain Account of Christian Perfection*, Epworth Press edition (London, 1970), p.5

[8] Wesley, John, *A Plain Account of Christian Perfection*, p.6

Wesley's approach to mission. Law's thoughts on the use of money and the importance of giving to the poor, Wesley wholeheartedly adopted.

All these ideas were helpful in the early years of the founding of the Holy Club in Oxford, which was a kind of working prototype for the later Methodist movement. The Holy Club had originally been founded by Charles Wesley who together with two other students, William Kirkham and William Morgan, met regularly to talk about religious books, and pledge themselves to engage in private prayer, and to frequently attend services of Holy Communion. On his return to Oxford, after serving for a short time as his father's curate, John took over the leadership of this group, and it was not long before the activities were extended to pastoral care and visiting prisoners in the Oxford Castle and the debtors' prison.

America

Following his father's death in 1735, John Wesley, together with his brother Charles, set out for America. His decision was influenced in part by the unsettlement of the family caused by his father's death, a desire to fulfil on his behalf an ambition to become a missionary, as well as an admiration for the ideals behind the founding of the new colony of Georgia. The concept of a new colony had come from Colonel Oglethorpe, who had undertaken on behalf of parliament a review of prisons. His report gave a scathing account of the poor treatment many prisoners were receiving, especially those in debtors' prisons. Oglethorpe believed that these people had no chance of a fresh start, or establishing a new business, or even finding employment. He therefore suggested the idea of setting up a new colony in honour of the King, where debtors could start a new life, together with Protestants fleeing from Europe. This new colony could then trade with Britain. Oglethorpe was so committed to the idea of a new colony that he offered to be its Governor. He sailed to the New World taking with him Charles Wesley as his secretary and

Embarking for America

John as a missionary to care for the local population and convert the native Americans.

The idealism of both brothers was soon crushed in Georgia. Charles returned early through ill health and a growing reluctance to deal with awkward congregations. John was falling in love with the 18 year-old Sophia Hopkey, but had not asked for her hand in marriage. However, he got so upset when she became engaged to a Mr Williamson that he excommunicated her. This resulted in a court case which had little chance of convicting Wesley but, when it was postponed for the sixth time, he lost patience and returned to England. His homecoming was overshadowed by a sense of failure, in that most things he had tried to do since leaving Oxford had not succeeded. He had set out for America with high hopes and a deep desire to do something worthwhile for God. Those hopes had evaporated and many of his efforts had ended in humiliation. Moreover his image as a man of action and integrity had been severely damaged by the court case.

However, his time in America was not a total failure. It had given him an opportunity to become conversant with Moravian Christians[9] and their simple way of life, and through them he had become familiar with the great German hymns. The Moravians had also reinforced his ideas on the importance of hymns and of singing the faith, both in public and in private. Secondly, he had produced the first hymn book to be printed in America, *A Collection of Psalms and Hymns* (1737).[10] This was the forerunner of many great hymn books that would follow.

Wesley's conversion

There is no doubt that for John Wesley this state of depression, failure and unhappiness, in which he returned home, was an important factor in his conversion. But there was also the positive influence on Wesley of Peter Böhler and the Moravians whom Wesley had long admired for their goodness, piety and simplicity of life. Böhler, a German Moravian Christian, was a deeply religious man who had undergone a religious conversion, and wished to encourage others to find the same faith he himself had discovered. Wesley, who had all the hurt and humiliation of America behind him and an increasing number of churches closing their doors to his preaching, was unusually receptive to Böhler's ideas.

During the four months between Wesley landing back in England and his conversion in London, his mind oscillated between faith and doubt, between a sense of God's presence and a feeling of being utterly deserted by God. It was the influence of Peter Böhler that shaped the course of Wesley's life during this period. Indeed the letter written by Böhler on 8 May as he was about to depart from Southampton for America greatly cheered Wesley. In the letter Böhler exhorted Wesley to 'Beware of the sin of unbelief; and if you

[9] Wesley, John, *Journal*, 24 February 1736
[10] This was the first hymn book to be published in America

have not conquered it yet see that you conquer it this day.'[11]

May 25, 1738 was an important day for John Wesley. That afternoon he went to St Paul's Cathedral where the choir sang an anthem based on one of Luther's favourite psalms, 'Out of the deep have I called unto thee, O Lord.' The combination of words and music made a great impression on Wesley. That evening he went 'unwillingly' to a religious society meeting in Aldersgate when as one of the members was reading Luther's *Preface to the Romans*, Wesley felt his 'heart strangely warmed. I felt that I did trust in Christ.'[12]

This was his conversion experience, but what difference did it make to his life? Dr James Rigg, writing a hundred years after Wesley's death, maintained that it changed him from being a high church man to an evangelical. Although there was an evangelical fervour about Wesley's subsequent mission, he still retained many of his high church practices such as regular celebration of Holy Communion and using the set prayers of the Church.

In later life, he even played down the importance of his conversion, and this was possibly because he had come to associate it with the Moravians from whom he wanted to distance himself.

What his conversion experience did give him was a deep sense of forgiveness. It also brought Christianity into a clear focus for him. It may not have been the great turning point in his life that some have claimed, but it is noticeable that from this time onwards his interests began to move away from Oxford and out into the wider world. His ill-received University Sermon of 1744, in which he criticized the members of the University for being lazy and immoral, marked the end of his effective association with Oxford[13], after which all his efforts were concentrated on preaching, travelling and building up his fledgling societies. In 1751 he formally severed his links with

[11] Wesley, John, *Letters*, edited by Frank Baker (Oxford, 1982). Letter from Peter Böhler, 8 May 1738

[12] Wesley, John, *Journal*, 24 May 1738

[13] Waller, Ralph, *John Wesley: A Personal Portrait*, (London, 2003), pp.69-71

Oxford when he resigned his Fellowship at Lincoln College upon marrying Mary Vazeille, a widow with three children.

Travelling and preaching

John Wesley, although only five feet two inches tall, was one of the most energetic leaders in the history of the Christian Church. Even St Paul's missionary journeys seem short in comparison with the journeys undertaken by John Wesley over a period of some fifty years. He travelled approximately 250,000 miles in his lifetime, and much of it on rutted tracks and poor roads. Before 1741 there were no turnpike roads in Britain north of the Humber, with the exception of a few miles around Whitehaven. It was left to local parishes to maintain the roads that ran through the parish, with the result that road surfaces were often of poor quality. Thirty years later, a good network of turnpike roads had been established throughout Britain; these linked together the major centres of the population. However even in the 1770s and 1780s once a traveller had moved off the main roads in the Yorkshire Dales, or the North Yorkshire Moors, the Lincolnshire Wolds or the Lancashire Hills, where Wesley was still travelling in old age, the roads were often rough and impassable by a horse and trap. On 23 May 1781, at the age of 77, Wesley wrote in his *Journal*: 'Having appointed to preach at Blackburn I was desired to take Kabb in on my way. But such a road sure no carriage ever went before: I was glad to quit it and use my own feet.'[14]

One of Wesley's followers, Bishop Asbury, boasted that in America he travelled 4,500 miles per year in order to preach the Gospel to a new generation of Methodists. Although such distances on horseback were great achievements, they seem less significant when compared with Wesley's annual travelling which averaged 6,000 miles a year.

Throughout his long life, Wesley rose at 4 a.m., preached at

[14] Wesley, John, *Journal*, 23 May 1781

John Wesley preaching on his father's tomb at Epworth

5 a.m. to those on their way to work, and then would often travel some sixty miles in the day preaching at three or four other places before the evening. From his early days he had learned to use his time carefully by reading while he walked. He further experimented and discovered that he could read books, write letters and prepare sermons whilst riding a horse. On the journey from Norwich to Yarmouth in 1765 he was not only able to read Isaac Watts' *The Improvement of the Mind,* but also produced a good critical account of the book before the end of his journey.

There is a widely held misapprehension that Wesley visited and preached in every village, and established a Methodist society in each place. There were large areas of the country where Methodism was practically unknown. Moreover Wesley was not constantly travelling around the country. In the dark winter months he tended to base his activities in London or Bristol, and make shorter journeys to towns within a hundred miles of London, such as Coventry or Dover. But once spring arrived, he would set out to Wales, Cornwall, the North East, and Ireland.

In the early years of his travelling, his plan was not to attempt to cover the whole country but to go to the areas where he felt he could make progress and where he would be given a hearing; these tended to be large areas of the population such as Newcastle, Bristol, and London. In these cities, where new industries were springing up, the population was less influenced by a restrictive parish system, dominated by the squire and the vicar. He tended to avoid the fashionable resorts, nor did he invest great energies in areas which were solely based on agriculture. These areas were close-knit communities where farm labourers were dependent for their livelihoods on being employed by the landowners, and thereby had less freedom than those in the larger cities. The rural areas which he often visited were those where agriculture was accompanied by industry, such as Epworth which had a flax industry, or Grimsby which had a prosperous fishing industry.

Wesley's approach to evangelism was very different to that of Whitefield who preached to large crowds, and then moved on hoping that others might build on his work. Wesley's strategy was to concentrate on particular centres and groups of people. His *Journal* clearly shows that in the first ten years as a travelling preacher, he concentrated on London, Bristol, and Newcastle where he began to expand his work westward up the Tyne valley and southwards into Weardale and Teasdale, and gradually down into the Leeds areas. It was not long before he extended his work into Cornwall and Staffordshire. There were groups of workers with whom he made rapid progress and these included the Cornish tin miners, the keelmen of Newcastle, the lead miners of Weardale and the fishing communities along the east coast of England.

Wesley's powers of endurance were remarkable. He continued to travel through the United Kingdom on horseback until he was 70 years old, when he began to use a small carriage or horse and trap. By the 1770s he was greatly assisted by the much improved turnpike road system which covered most of the country. Indeed he probably paid more turnpike road tolls than any other person in England, although he himself had opposed these new charges, not because he objected to paying himself, but on the grounds that they limited travel for poor people who could not afford the tolls.

The perils of travelling in the eighteenth century were not limited to poor roads, dangerous ferries, and runaway horses, but there was also a threat from highwaymen, which Wesley and Whitefield occasionally faced.

When he was 87 he was still preaching in Yorkshire, Lincolnshire, and Norfolk. On 13 October 1790 he preached in Norwich, and the congregation could not be contained within the building. Old age did not prevent him travelling to Newcastle, Cornwall, and Ireland all of which held a special place in his affections.

It is not easy to discern his essential message from the numerous published sermons that survive from the 45,000 that he preached. However, Wesley clearly set out the heart of his message in a conversation with a previously wealthy man who had lost his fortune and tried to take his own life.[15] When asked what was the message he preached, Wesley replied that it was to make people virtuous and happy, at ease with themselves, and useful to others. He wished to lead those who would listen to heaven, and to God, and to Jesus. Wesley went on to emphasize that he preached the religion of love and kindness that he found in the Gospels. It was his hope that his hearers would come to enjoy God and overcome the fear of death.

Care for people

Wesley's success was due not only to his effective preaching and extensive travelling, but also to a genuine love of people and the

[15] Wesley, John, *Works*, 3rd edition, 14 vols, (London, 1872), Vol.VIII:9-10

deep social concern that he expressed towards those in need. He was one of the few people in the eighteenth century who did not distinguish between the deserving and the undeserving poor. He saw all poor people as those in need. Early on in his ministry, in the bitter winter of 1741, he came across forty destitute agricultural workers and their families, unable to work because the ground was so frozen.[16] He took collections for them and later collected clothes for them. He instigated a system whereby every Methodist would donate one penny a week to help feed and clothe the poor. As an old man well over the age of 80, we find him in London going from house to house on four successive bitter winter days, collecting money for poor unemployed building workers and their families. It had snowed heavily and the snow had frozen, but was thawing and the slush and icy-cold water was coming over the top of his boots, but he did not stop until he had collected £200 with which to feed and clothe the poor. He wrote in his *Journal* that he did not like doing this but, if he did not do it, nobody else would.[17]

He set up free dispensaries in London, Bristol and Newcastle so that poor people could have good medical treatment. He studied medicine and also employed a chemist and a surgeon to help. His first patient was a London weaver called William Kirkman who came to Wesley with a bad cough. When Wesley asked him how long he had had the cough, the man replied, 'Seventy years.' Wesley was fearful of failing with the first patient in case the others would lose confidence in him, but he gave him some cough mixture and told Kirkman that it would do him no harm and might do him some good. Within three days the cough was cured.[18]

His social concern was vast and extended to schools for children and adults, alms houses, knitting cooperatives, prison visiting and reform, establishing loan funds, redistribution of clothing, and generous personal giving. He once said that he would not need an

[16] Wesley, John, *Journal*, 21 January 1740
[17] Wesley, John, *Journal*, 5 December 1785
[18] Wesley, John, *Letters*, Letter to Vincent Perronet, December 1748

executor, as he intended to be his own executor and give away his money while he was still alive. Wesley had a deep interest in people and their material well-being, especially the poor and children. He preached the love of God and brought faith and good works together in a manner in which few evangelical leaders have managed to do before or since. His spirituality always found expression in material things and in doing good.

Wesley changed the lives of individuals. He gave poor people a sense of value and worth and reminded them that they too were children of God. Through his efforts, keelmen and miners, prostitutes and prisoners, sailors and smugglers, became devout people with a purpose in life and a value in society. The Methodist movement that he established, saved men and women from poverty, aimlessness and degradation, and gave them a glimpse of heaven. It taught them to hope in the hereafter, but it also made them stable parents, good citizens, and caring people.

Ralph Waller

2
Geographical publisher and traveller

Geographical publisher

The appeal of Wesley's religious approach meant that his followers were in the main from the lower, less educated, classes. His concern to support them included significant efforts to publish intelligible and affordable material so they could improve their lot through reading and knowledge. He established many schools and reading classes for his followers and lay preachers, playing a key role in supervising their curricula and reading lists. It soon became obvious that in many areas suitable affordable books were simply not available, and this led to the establishment of his printing works where such books could be produced cheaply. He became the leading writer, editor, publisher and printer in the country with a total of over 300 publications to his name. By selling his books cheaply under his name, and with encouragement of his local preachers, he was successful in extending guided education, including scientific and geographical subjects, to significant numbers.

Apart from his hymns, sermons and addresses of various types, the best known of his publications was his *Journal* which is used as the basis of the majority of this book. This was published in cheap instalments and represents a carefully considered re-working of his travels and experiences. It is therefore far from being a spontaneous account. Among the various books he published were three related to science and geography. It may at first seem surprising that he strayed into such topics, but he was insistent that an all-round education which included the sciences was important. In his address to clergy he stresses the importance of being 'a tolerable master of the sciences' and states:

Asking directions

'Some knowledge of the sciences, is to say the least, equally expedient *(compared to other academic disciplines)*. Nay, may we not say, that the knowledge of one, (whether art or science,) although now quite unfashionable, is even necessary next, and in order to, the knowledge of the Scripture itself?'

Later in his life he supplemented his *Journal* with *The Arminian Magazine*. This was issued in separate monthly instalments to provide an acceptable alternative to existing Calvinistic magazines and reached an annual circulation over 7,000 copies. The family and communal fellowship encouraged lending it to others so the number of readers was far in excess of this number. The magazine also included scientific and geographical articles with travel accounts and extracts from his condensed books.

There is also one apparently original scientific, meteorologically related, article published by Wesley: 'A question concerning dew on coach glasses.'

> 'When two or three persons are in a coach, and draw up the glasses, it is observed they become so covered with dew, that we can scarce see through them; but when that is once wiped off, there is no more dew gathered upon the glass, but it continues transparent. You oblige your readers with the reason for this phenomenon.
>
> THE ANSWER
>
> The reason is, that in comparison of the moist vapours that come from the persons in the coach, the glass is cold and condenses them, remaining colder longer than any other part of the coach; as we find in damp weather, that marble will become wet by condensing the moisture of the air. Then, by degrees, the glass, partaking of the warmth of the persons in the coach, is no longer able to condense the floating vapours into water. The proof of this is plain by letting down the glass into its place, because there it cools, and then being brought up again condenses the vapour and gathers the dew; without which it would not condense the vapour, though in many hours' travelling.'

Wesley has in this short piece demonstrated his understanding of the physics and meteorology of the condensation process. It was the careful determination of the temperature of glass upon which water condenses (the dew point) that later formed the basis of air humidity measurement used by John Dalton in the early 1800s.

Wesley published three books on scientifically related subjects. Firstly, *Primitive Physic*. This is an extensive collection of home remedies and general medical advice covering a wide range of illnesses. Although criticised, it met with popular acclaim and went through over thirty editions and continued to be published well after his death. His concern for the health of his followers was also demonstrated in the medical support and clinics that were provided.

Secondly, *The Desideratum* on electricity. Wesley was impressed with demonstrations of electrical experiments and with the potential of the use of electricity for medical treatment. After he had read letters by Benjamin Franklin on electricity, he published a condensed version of some of them in his *Journal* for 7 February 1753. He decided that the subject area was important enough for his followers to be made aware of it and in 1760 published *The Desideratum* in which he 'endeavoured to make electricity plain' and covered both electrical experiments and medical applications. He was convinced of the curative power of electricity and purchased 'electrical machines' to use for treatment.

Thirdly, *A Survey of the Wisdom of God in the Creation, or a Compendium of Natural Philosophy*. This represents Wesley's main contribution to public understanding of physical geography. Like many other Wesley articles and books the information is drawn from publications by other authors and reproduced in condensed form, sometimes with little or no acknowledgement to their origin. For the original version he stated that 'the text is in great measure translated from the Latin work of John Francis Buddeus ... but I have found occasion to retrench, enlarge or alter every chapter and almost every section.' He also used other works for additional notes.

His aim was to publish a book to describe but not necessarily explain scientific phenomena:

> 'In the plainest dress, simply and nakedly expressed, in the most clear, easy and intelligible manner that the nature of things would allow: particularly free from the jargon of mathematics, which is mere heathen Greek to common readers ... directed to its right end: not barely to entertain an idle barren curiosity but to display the invisible things of God, his power, wisdom and goodness ... I endeavour throughout, not to account for things, but only describe them ... The facts lie within reach of our senses and understanding; the causes are more remote. In many case we cannot know; ... God has done his Works, that we may admire and adore: But we cannot search them out to perfection.'

The original work effectively comprised a scientific encyclopaedia collated from a range of sources. In addition to the formal book version information was also made more widely available by printing extracts in *The Arminian Magazine.*

It is, however, not until the third edition that the most relevant geography-related material is included. This includes a compressed form of the book *Contemplation of Nature* by Charles Bonnet. The latter sections of these editions also include a summary of Deuten's *Inquiry of the origin of the discoveries attributed to the Moderns* which, amongst a vast range of topics, includes some early studies on thunderstorms and rivers.

The main geography material appears within three chapters:

Chapter 1: *Of Earth and Water.* This includes sections on the formation and structure of the earth; mountains and their heights; rivers, their types and origin (here again the principle, later confirmed by John Dalton, that the water from all rivers is supplied by evaporation and rainfall processes is included); the seas and tides; subterraneous trees, and bogs.

Chapter 2: [*It carries no title.*] This covers geological and geomorphological topics including detailed descriptions of various volcanoes (including Etna, Vesuvius and Hecla) and earthquakes (including related writings by Pliny, and details from Port Royal in Jamaica and Lima).

Chapter 3: *Of Meteors*. This chapter first describes many aspects of the science of meteorology including mists, clouds, dew, rain, snow, hail, the rainbow, haloes and mock suns. It then continues with descriptions of an eclectic mix of natural history and electrical phenomena including the aurora. The final part covers winds and climates in various locations, and extreme events such as water spouts, and ends with 'reflections on the usefulness of the atmosphere to the whole globe.'

The majority of the rest of the book concentrates on Wesley's own observations and comments, mainly originating during his extensive journeys.

Wesley's travels

After his return from America, having survived very stormy passages in both directions, John Wesley included extensive preaching expeditions in his hectic annual work schedule. Roads over this period were gradually improving with increasing numbers of 'turnpike' [*toll*] roads becoming available. Many journeys were also made by water, and events from his *Journal* give many examples of the real dangers that could sometimes be involved.

By the 1720s road maps were becoming available in published form. He did, however, frequently resort to the use of local guides, especially for more remote areas. By the end of the century road maps were becoming more comparable with modern maps, and indeed the first British Ordnance Survey maps were published in the early nineteenth century.

For most of his career journeys were generally made on horseback, with Wesley often avidly reading at the same time as riding. He frequently got soaked and frozen, but, as will be seen later, it was hot weather that often seemed to cause him most distress. Not being sympathetic to modern clothing he generally stuck to standard wear of the time which quickly became sodden in wet weather. He does, however, mention on one occasion of grudgingly agreeing to use the protection a modern 'sou'wester' with obvious success:

> ***Journal*, 16 Jan 1764:** I rode to High-Wycombe, and preached to a more numerous and serious congregation than I ever saw there before ... A large number was present at five in the morning: But my face and gums were so swelled I could hardly speak. After I took horse, they grew worse and worse, till it began to rain. I was then persuaded to put on an oil-case hood, which (the wind being very high) kept rubbing continually on my cheek, till both the pain and the swelling were gone.

As with comments about adverse weather, he became more aware about bad roads as he grew older and, despite resorting ever more frequently to wheeled transport, travel inevitably became more of a strain. A few examples from his *Journal* will give a feel for the types of journeys he had to contend with:

> ***Journal*, 31 Jul 1764:** We set out for Glamorganshire, and rode up and down steep stony mountains for about five hours, to Larn. Having procured a pretty ready passage there, we went on to Lansteffan-Ferry, where we were in some danger of being swallowed up in the mud before we could reach the water. Between one and two we reached Kidwelly, having been more than seven hours of horseback, by which time we could have rode to Carmarthen with more ease to both man and beast. I have, therefore, taken my leave of these ferries; considering we save no time by crossing them, (not even when we have a ready passage,) and so have all the trouble, danger, and expense, clear gains. I wonder that any man of common sense, who has once made the experiment, should ever ride from Pembroke to Swansea any other way than by Carmarthen.

An honest man at Kidwelly told us there was no difficulty in riding the sands, so we rode on. In ten minutes one overtook us who used to guide persons over them; and it was well he did, or in all probability we had been swallowed up. The whole sands are at least ten miles over, with many streams of quicksands intermixed. But our guide was thoroughly acquainted with them, and with the road on the other side. By his help, between five and six, we came well tired to Oxwych in Gower.

***Journal*, 19 Jun 1787:** We went on through horrible roads to Newry. I wonder any should be so stupid as to prefer the Irish roads to the English. The huge unbroken stones, of which they are generally made, are enough to break any carriage in pieces. No, there is nothing equal to good English gravel, both for horses, carriages and travellers.

***Journal*, 21-23 Apr 1788:** [*An example of Wesley's use of the ironic opposite.*] I went on, through miserable roads, to Blackburn ... Through equally good roads we got on to Padiham . . . from hence we went, through still more wonderful roads, to Haslingden. They were sufficient to lame any horses, and shake any carriage to pieces. – N.B. I will never attempt to travel these roads again, till they are effectually mended!

However, on a more positive note:

***Journal*, 15 Jun 1770:** I was agreeably surprised to find the whole road from Thirsk to Stokesley, which used to be extremely bad, better than most turnpikes. The gentlemen had exerted themselves, and raised money enough to mend it effectually. So they have done for several hundred miles in Scotland, and throughout all Connaught in Ireland; and so they undoubtedly might throughout all England, without saddling the poor people with the vile imposition of turnpikes for ever.

> ***Journal*, 13 May 1788:** To-day we went on through lovely roads to Dumfries. Indeed all the roads are wonderfully mended since I last travelled this way. Dumfries is beautifully situated; but as to wood and water, and gently rising hills etc., is, I think, the neatest as well as most civilized town that I have seen in the kingdom.

Subsequent chapters contain many examples of travel by water, both sea crossings and coastal journeys, but despite the obvious dangers, Wesley's attitude was still scathing of the reluctance of captains to take risks:

> ***Journal*, 28 Feb, 1748:** I never knew men make such poor, lame excuses, as these Captains did for not sailing. It put me in mind of the epigram
>
> *There are, if rightly I methink*
> *Five causes why a man should drink;*
>
> which, with a little alteration, would just suit them:
>
> *There are, unless my memory fail,*
> *Five causes why we should not sail:*
> *The fog is thick; the wind is high;*
> *It rains; or may do by-and-by;*
> *Or – any other reason why.*

3
Weather reporter

Introduction

As John Wesley was on the move a great deal of the time, it is not surprising that his *Journal* and letters often include comments about the weather conditions he encountered. Covering such an extensive area, these reports are treated separately from other topics in physical geography he wrote about.

His first written comments are as early as 1724 in a letter to his mother from Oxford and already show his interest both in meteorology and in the natural environment which continued throughout his life:

> **Letter, 18 Dec 1724:** 'We still have very warm weather at Oxford; and a gentleman, now in the room with me, says that several of the flowers in his father's garden, who lives in town, are blown as if it were Spring.'

Wesley's letters include regular use of meteorological similes to stress a point:

> **Letter, Leeds, 7 Oct 1749:** [*This letter concerning his lost love is considered among the most tragic he ever wrote.*] ... But we were soon after torn asunder by a whirlwind. In a few months the storm was over ... The waves soon arose again ...

> **Letter, Dublin, 12 Apr 1750:** I doubt you are in a great deal more danger from honour than from dishonour. So it is with me. I always find there is most hazard in sailing upon smooth water.

When the winds blow and the seas rage, even the sleepers will rise and call upon God.

Letter, Epworth, 16 Apr 1752: ... I have seen more make shipwreck of the faith in a calm than in a storm. We are apt in sunshiny weather to lie down and sleep; and who can tell what may be done before we awake?

Letter, Birstall, 28 May 1753: ... I believe the harvest has not been so plenteous for many years as is now in all the North of England; but the labourers are few.

Letter, Bristol, 19 Sep 1773: It is certainly most profitable for us to have a variety of seasons. We could not bear either to be constantly in storms or constantly in a calm.

Letter, London, 26 Nov 1786: In the time of Dr Jonathan Edwards there were several gracious showers in New England, but there were large intermissions between one and another; whereas with us there has been no intermission at all for seven-and-forty years.

Letter, Parkgate, 25 Apr 1790: ...You keep your love and you keep your integrity even among weathercocks. But who was it that turned Dr Coke from east to west?

The eighteenth century saw a wide range of climates with some notable meteorological extremes. A brief summary of the British conditions over Wesley's travelling life from 1730-90 is given below. (For more information on the weather conditions in the 1700s see John Kington's book *Climate and Weather*, Collins, 2010).

The period began with predominating westerly winds and is one of the warmest decades on record with no cold winters except for 1731. This was in marked contrast to the 1740s when the warm westerly flows diminished. This decade began with the very cold winter of 1740 precipitating mass starvation in Ireland. The early years were also exceptionally dry. The cold conditions continued into the next decade with the climax of the 'Little Ice Age' occurring around 1750. The wet 1760s ended with an abrupt change in atmospheric circulation and the end of the 'Little Ice Age' in 1770. This was then followed by a brief warm interlude from 1773-81 with excellent wine harvests being recorded. Conditions were adversely affected by the volcanic ash cloud of 1782 and farming was also hit by the prolonged drought of 1784-5. After further bad winters the 1780 decade culminated with the extreme winter of 1788-9 which was famed for its frost fairs on the Thames but associated deprivation of the population.

Wesley only travelled, and therefore experienced, weather conditions in Britain and the near continent with the one exception of his time in Georgia in the 1730s. The conditions which made the most impression on him, and to which he refers in his *Journal* on several occasions, are the high temperatures and the violent thunderstorms.

It is indeed interesting to note that, whereas the modern long-term average maximum temperature in Savannah (as published by http://www.weather.com) of 22°C (71°F) is much the same as in England, the extreme Savannah maximum of 41°C (105°F) is significantly higher than those for England. If Wesley experienced temperatures approaching this level it is not surprising they made a lasting impression on him.

The most severe British storm of his lifetime was in December 1703, shortly after his birth. One of the greatest historically recorded storm surge disasters ever was in December 1717 when 11,000 died on the continental side of the North Sea.

In order to get the best picture of the weather he endured during his life and his response to the conditions, the following sections give selected extracts from Wesley's *Journal* and letters in chronological order, treating each weather type separately. Such detailed comments are of potential use to those involved with building up descriptions of local conditions for use in climate change and social studies.

One must of course bear in mind that, except for reported spectacular events, he only tended to describe weather conditions which he experienced which actually affected his work. Consequently, reports are normally confined to periods when he was travelling. So, for instance, some of the extremely cold periods of the century do not get significant mention.

Temperature extremes

Wesley was generally able to endure cold conditions without many complaints but, later in life, when often preaching outside, the elements seemed to have more impact on him as shown in many of these examples:

> ***Journal*, 19 Mar 1770:** I rode to Craidley. Here also a multitude of people obliged me to stand abroad although the North wind whistled round my head ... At six I began at Dudley. The air was as cold as I almost ever felt. But I trust God warmed many hearts.

> **Letter, Deptford, 20 Nov 1775:** We had hardly any frost last winter; perhaps you will have little more the ensuing. I am commonly more pinched by the November cold than by that which comes after Christmas.

> **Letter, London, 21 Dec 1775:** I am surprised that one who has passed a winter in Scotland should complain of cold in Wales. It is not a good sign.

> ***Journal*, 19 Mar 1784:** As we rode back to Bengeworth, the cold was so intense, that it had an effect I never felt before, – it

made me downright sick. However I went on and preached in the church at eleven and in the evening at Worcester.

***Journal*, 29 Mar 1784:** We went to Lane-End, a village two or three miles from Newcastle-under-Lyme. It was still piercingly cold; but the preaching house would not hold a fourth part of the people. So I preached in the open air; the moon giving us a clear light but not much heat. [*This is one occasion where Wesley's scientific understanding lets him down as there is actually no heat received from the moon, and indeed a bright moon means a clear sky and greater heat loss from the surface leading to increased cooling!*]

Letter, Bristol, 16 Sep 1789: It gives me pleasure to hear that you have so much resolution that you go to bed at ten and rise at four o'clock. Let not the increase of cold affright you from your purposes.

When advising on the design of a chapel Wesley was concerned not to chill his congregation:

Letter, London, 12 Feb 1769: A preaching-house cannot be too light or too airy. Therefore your windows must be large. And let there be sashes, opening downward; otherwise air coming in would give the people cold.

Freezing conditions

Wesley encountered all the rigours of a Georgia winter during his time in America as illustrated by this extract:

***Journal*, 22 Dec 1736:** There was not time to walk back to Savannah before night … The ground was as wet as our clothes, which (it being a sharp frost) were soon froze together; however,

I slept till six in the morning. There fell a heavy dew in the night which covered us as white as snow.

On return to England there were many occasions when Wesley reported on freezing conditions. For example:

***Journal*, 19 Feb 1747:** The frost was not so sharp so we had little difficulty till we came to Haxey-Car; but here the ice which covered the dykes, and a great part of the Common, would not bear, nor readily break; nor did we know (there being no track of man or beast) what parts of the dykes were fordable.

***Journal*, 21 Nov 1748:** I set out for Leigh, in Essex. It had rained hard in the former part of the night, which was succeeded by a sharp frost; so that most of the road was like glass; and the north-east wind set just in our face.

***Journal*, 30 Jan 1751:** I set out early [*to Oxford*], in a severe frost, with the north wind full in my face. The roads were so slippery, that it was scarce possible for our horses to keep their feet. Indeed one of them could not; but fell on his head, and cut it terribly.

During the same cold spell Wesley had an accident in the slippery conditions:

***Journal*, 10 Mar 1751:** On the middle of London-Bridge, both my feet slipped on the ice, and I fell with great force, the bone in my ankle lighting on the top of a stone. However I got on, with some help, to the chapel, being resolved not to disappoint the people.

Letter, Bristol, 5 Jan 1754: We are (quite contrary to my judgement, but our friends would have it so) in a cold, bleak place, and in a very cold house. If the hot well water makes amends for this it is well.

The following extract contains a reference to temperature values but it is not clear whether he is reporting his own observations or merely quoting temperature reports he has heard or read about. In this case the latter possibility seems more likely. This reference occurs right at the beginning of one of his *Journal* sections covering the period January 1776 to August 1779:

***Journal*, 17 Jan 1776:** For some days before the frost broke up, it was observed, by means of the thermometer, that the cold was several degrees more intense than that in the year 1741. But God then removed the cup from us, by a gentle, gradual thaw.

The impact of the great freeze of the winter of 1788 is clearly described:

***Journal*, 15 Dec 1788:** In the evening I preached at Miss Teulon's school in Highgate. I think it was the coldest night I ever remember. The house stood on the edge of the hill, and the east wind set full in the window. I counted eleven, twelve, one, and was then obliged to dress, the cramp becoming more and more violent.

***Journal*, 25 Dec 1788:** (Being *Christmas-day*) We began the service as usual at four in the new chapel. Notwithstanding the severe frost which had now lasted a month, the congregation was uncommonly large. [*How many 4 a.m. Christmas services would be well-attended nowadays?*]

***Journal*, 30 Dec 1788:** The frost continued severe, yet the congregation was uncommonly large in the evening, and God warmed many hearts.

***Journal*, 4 Jan 1789:** Although the extreme severity of the weather kept many tender people away; yet we had a large congregation in the evening to renew their covenant with God.

Snowy conditions

As he frequently travelled large distances during the winter months Wesley regularly encountered snowy conditions, which he generally intrepidly ploughed on through. However, some were especially difficult:

***Journal*, 23 Feb 1745:** We found the roads abundantly worse than they had been the day before; not only because the snows were deeper, which made the causeways in many places impassable, (and turnpike roads were not known in these parts of England till some years after) but likewise because the hard frost, succeeding the thaw, had made all the ground like glass. We were often obliged to walk, it being impossible to ride and our horses several times fell down while we were leading them, but not once while we were riding them, during the whole journey. It was past eight before we got to Gateshead-Fell, which appeared a great pathless waste of white. The snow filling up and covering all the roads, we were at a loss how to proceed; when an honest man of Newcastle overtook and guided us safe into the town.

Many a rough journey have I had before, but one like this I never had; between wind, and hail, and rain, and ice, and snow, and driving sleet, and piercing cold.

***Journal*, 20 Feb 1746:** Before we came to Aldridge-Heath, the rain changed into snow, which the northerly wind drove full in our faces, and crusted over us from head to foot in less than an hour's time. We inquired of one who lived at the entrance to the moors, which was our best way to Stafford. 'Sir,' he said, "tis a thousand pound to a penny, that you do not come there today.

Why, 'tis four long miles to the far side of this common; and in a clear day, I am not sure to go right across it; And now all the roads are covered with snow; and it snows so, that you cannot see before you.' However, we went on, and I believe did not go ten yards out of the way till we came into Stafford.

Snowy conditions at Gateshead-Fell

***Journal*, 17 Feb 1747:** We set out as soon as it was light; but it was really hard work to get forward; for the frost would not well bear or break: And the untracked snow covering all the roads, we had much ado to keep all our horses on their feet. Meantime the wind rose higher and higher, till it was ready to overturn both man and beast. However, after a short bait at Bugden, we pushed on, and were met in the middle of an open field with so violent a storm of rain and hail, as we had not had before. It drove through our coats, great and small, boots and every thing, and yet froze as it fell, even upon our eyebrows; so that we had scarce either strength or motion left, when we came into our inn at Stilton.

We now gave up our hopes of reaching Grantham, the snow falling faster and faster. However, we took the advantage of a fair blast to set out, and made the best of our way to Stamford-Heath. But here a new difficulty arose, from the snow lying in large drifts. Sometimes horses and man were well nigh swallowed up. Yet in less than an hour we were brought safe to Stamford. Being willing to get as far as we could, we made but a short stop here; and about sunset came, cold and weary, yet well, to a little town called Brig-Casterton.

~ **18 Feb:** Our servant came up and said, 'Sir, there is no travelling to-day. Such a quantity of snow has fallen in the night, that the roads are quite filled up.' I told him 'At least we can walk twenty miles a day, with our horses in our hands.' So in the name of God we set out. The north-east wind was piercing as a sword, and had driven the snow into such uneven heaps, that the main road was unpassable. However we kept on, a-foot or on horseback, till we came to the White Lion at Grantham.

***Journal*, 1 Dec 1747:** About noon we reached Stockbridge. The rain then changed to snow. Seeing no prospect of fair weather, after resting a while we set out in the midst of the storm. It blew such a hurricane, as I have scarce known in England, and that full in our teeth, so that our horses reeled to and fro, and had much ado to keep their feet. The snow likewise drove vehemently in our faces, in riding over the open Downs, where, for several miles, there was neither house, nor tree, nor shrub to shelter, that it was hard labour to get forward. But in about an hour, the sky cleared up, and we rode comfortable to Salisbury.

***Journal*, 16 Feb 1748:** We passed through Chepstow soon after sunset, and pushed on; though it grew dark, and the untracked snow lay thick upon the ground. About eight we reached the Star, a good, though small inn, five long miles from Chepstow.

~ **17 Feb:** It snowed all night. We set out before day; but found it bad travelling, there being no path to be seen, neither footsteps of man or beast. However, in four or five hours we reached Abergavenny; and Brecknock before three in the afternoon.

Our landlady here almost forced us to take a guide. And it was extremely well we did; for the snow had so entirely covered the roads, that our guide himself mistook the way more than once. So that if he had not been with us, we should, without doubt, have lodged upon the mountains.

Travelling further North later in the month Wesley makes an unexpected observation on the distribution of the winter snowfields:

***Journal*, 22 Feb 1748:** I preached again at Garth, and on *Monday*, 22, at five in the morning. A little before sun-rise we took horse, it being a clear sharp frost. We had waited four days in the hopes that the snow would melt, fearing the drifts of it would lie deep upon the mountains, particularly as we journeyed northward; but quite contrary to our expectation, the farther northward we went, the less snow we found, so that it scarce hindered us after the first day.

***Journal*, 22 Nov 1759:** I took horse between six and seven, in one of the coldest mornings I ever remember. We reached St Alban's without much difficulty; but then the roads were all covered with snow. However, there was a beaten path, though slippery enough, till we turned into the by-road at Sundon. What we could have done there I cannot tell for the snow lay deeper and deeper, had not a waggon gone awhile before us, and marked the way for six miles, to Mr Cole's gate.

~ **23 Nov:** The roads were so extremely slippery, it was with much difficulty we reached Bedford.

Travelling in the Scottish Highlands Wesley encountered problems with snow even in late April:

> ***Journal*, 25 Apr 1770:** We went on to the Castle of Blair. The mountains for the next twenty miles, were much higher, and covered with snow. In the evening we came to Dalwhinny, the dearest inn I have met with in North-Britain. In the morning we were informed so much snow had fallen in the night that we could get no further. And indeed, three young women, attempting to cross the mountain to Blair, were swallowed up in the snow. However, we resolved, with God's help, to go as far as we could. But about noon we were at a full stop: The snow, driving together on the top of the mountain, had quite blocked up the road. We dismounted, and, striking out of the road warily, sometimes to the left, sometimes to the right, with many stumbles, but no hurt, we got to Dalmagarry, and before sunset to Inverness.

Hail

There are several specific mentions of severe hail storms within the *Journal*. However, by the time reports of past storms reached Wesley significant exaggeration seems to have taken place:

> ***Journal*, 17 Mar 1756:** I learned the particulars of that surprising storm which was here the year before last. It began near Cheltenham, on June 14, 1754, and passed on over Coleford, in a line about three miles broad. It was rain mixed with hail. The hail broke all the windows it had access to, stripped all the trees both of fruit and leaves, and destroyed every green thing. Many of the stones were as large as hen-eggs: some were fourteen or fifteen inches round. The rain occasioned such a torrent of water in the street as bore away man and beast. A mile or two farther, it joined with the waters of a mill-dam; which it broke down, and

carried away several houses. How frequent would accidents of this kind be, if chance, not God, governed the world!

***Journal*, 2 Nov 1763:** On Friday, August 19, a gentleman who was at Lisbon during the great earthquake, walking with his friend near Brighthelmstone, in Sussex, and looking south-west toward the sea, cried out, 'God grant the wind may rise; otherwise we shall have an earthquake quickly. Just so the clouds whirled to and fro, and so the sky looked [*like*] that day at Lisbon.' Presently the wind did rise, and brought an impetuous storm of rain and large hail. Some of the hail-stones were larger than hen-eggs. It moved in a line about four miles broad, making strange havoc, as it passed quite over the land, till it fell into the river not far from Sheerness. And whenever it passed it left an hot sulphurous steam, such as almost suffocated those it reached.

***Journal*, 21 Mar 1779:** Just at the time of preaching, at Bromwich-Heath, began such a storm as that which ushered in the year. Yet as no house could contain the people I was constrained to stand in the court-yard. For a moment I was afraid of the tiles falling on the people; but they regarded nothing but the word. As I concluded, we had a furious shower of hail; Hitherto could the prince of the power of the air go; but no farther.

Hot weather

John Wesley was more upset by hot weather than by even the most extreme cold conditions. This applied not only to exposure outside but also to hot conditions and the effect of sun shining on him while he was indoors:

***Journal*, 14 Jun 1741:** In the afternoon we returned to Markfield. The church was so excessive hot, (being crowded in every corner,) that I could not, without difficulty, read the Evening service.

***Journal*, 12 Jun 1752:** I rode to Bolton. So hot a day as this, I do not remember to have felt in England. The congregation seemed to forget the heat, though the Room was like an oven. For it was a comfortable hour: God refreshing many souls with the multitude of peace.

***Journal*, 11 May 1777:** I experienced a second time what one calls *febris ex insolatione* [a fever produced by an incautious exposure to the sun – *editor's comment in a later edition*]. The day was cold; but the sun shone warm on my back, as I sat in the window. In less than half an hour I began to shiver, and soon after had a strong fit of an ague. I directly lay down between blankets, and drank largely of warm lemonade. In ten minutes the hot fit came on, and quickly after I fell asleep. Having slept half an hour, I rose up and preached.

Most of Wesley's very many hot weather experiences that he describes are of course associated with travel or outside preaching. There are many comments about conditions being as hot as he ever remembered in England, or, as we will see later, in Georgia. One must obviously not take all of these literally, as memory of past weather can play tricks over time. However, his comments about the length of current periods of hot weather can be taken more literally. As would be expected there are more complaints in general about weather conditions as he gets older:

***Journal*, 31 May 1753:** I rode through a delightful vale to General Wood, near Todmorden. The sun was burning hot; but they set up a little tent for me … The afternoon was the hottest I ever remember in England: So by the time we came to Bolton, I was fit for nothing but to lie down.

Wesley was always interested in supplementing his personal experiences for his *Journal* with additional relevant information gleaned from people he met. This next extract is an excellent example of this. It also shows his appreciation of the importance, in the days when thermometer calibrations were not very precise, of using the same instrument to take temperatures observations in two different locations if the comparisons are going to be believed:

> ***Journal*, 11 Jul 1757:** This and the next three days were the hottest I ever knew in England. A gentleman who formally traded in Guinea, assured me, that the spirits in his thermometer (the same he had when abroad) rose as high as they did within a few degrees of the line.

> ***Journal*, 17 Mar 1766:** I rode to Stroud, the sun shining as in May. Seventeen such days in the beginning of March, I suppose, few man have known. But on *Tuesday* the weather entirely changed; and the piercing north-east wind, this and the two following days, was scarce supportable.

> ***Journal*, 5 Aug 1769:** In the evening I preached at Bradford, to an extremely crowded audience: The heat was hardly supportable. Such a day I had seldom, if ever, known in England. It was nearly as hot at four in the morning, but the rain began before five, and in three or four hours quite cooled the air.

> ***Journal*, 4 Jun 1775:** The sun was intensely hot, as it was on Monday and Tuesday. Six such days together, I was informed, have not been in Ireland for several years.

> ***Journal*, 15 Mar 1779:** I began my tour through England and Scotland; the lovely weather continuing, such as the oldest man alive has not seen before, for January, February, and half of March.

Wesley sometimes mentions temperature values overseas he knows of to compare them with what he is experiencing at the time:

> ***Journal*, 29 Aug 1779:** [*At Bristol*] It was one of the hottest days I have known in England. The thermometer rose to eighty degrees; as high as it usually rises in Jamaica.

The following entry appears to be the only mention of John Wesley actually having access to his own thermometer in London as well as again being aware of typical temperature conditions overseas:

> ***Journal*, 9 Jul 1783:** These four days, *Tuesday, Wednesday, Thursday,* and *Friday,* were as hot as the midsummer days in Jamaica. The summer heat in Jamaica usually raises the thermometer to about eighty degrees. The quicksilver in my thermometer now rose to eighty-two.

> ***Journal*, 7 Jul 1784:** In the afternoon the heat was scarce supportable, and it seemed to increase every hour; but between two and three in the morning came a violent storm, followed by uncommon thunder, and a flood of rain; this entirely cleared the air, and, ceasing just as we set out, left us a pleasant journey to Wakefield.

> ***Journal*, 14 Jul 1784:** The hottest day we have had this summer. We reached Tenby soon after one.

The Georgia connection

Wesley's traumatic time in Georgia made a permanent impact in many ways. From a meteorological standpoint the high summer temperatures and other weather conditions he experienced, such as these below, were obviously fixed in his memory:

***Journal*, 10 May 1736:** I began visiting my parishioners in order, from house to house: for which I set apart (the time when they cannot work because of the heat, viz.) from twelve to three in the afternoon.

***Journal*, 2 Dec 1737:** Thunder and lightning are expected almost every day in May, June, July, and August. They are very terrible, especially to a stranger. During those months from ten in the morning to four in the afternoon, the sun is extremely scorching. But the sea-breeze generally blows from ten till three or four. The winter is nearly of the same length as in England. But the mid-day sun is always warm, even when the mornings and evenings are very sharp, and the nights piercing cold.

In later years he refers back to these conditions, such as in these entries concerning hot weather:

***Journal*, 16 May 1758:** We rode to Lurgan …This was the hottest day I ever felt in Ireland: Near as hot as any I remember in Georgia.

***Journal*, 26 May 1762:** We took horse at four, to enjoy the cool of the morning. At seven the sun was warm enough: verily I think as warm as in Georgia. We could not have borne it, but the wind was in our face. However, in the afternoon we got well to Galway.

~ 27 May: We had another Georgian day; but having the wind again full in our face, after riding about fifty English miles, we got to Ennis in the afternoon.

***Journal*, 5 Aug 1770:** [*At St Albans*] *Sunday* and for five or six days this week, the heat was as great as I remember in Georgia.

***Journal*, 12 Aug 1773:** We came to Collumpton. For five or six days, I think, the weather has been as hot as it is in Georgia.

***Journal*, 1 Jul 1779:** This was the first of eighteen or twenty days full as hot as any I remember in Georgia; and yet the season is remarkably healthy. I preached at Beverly at noon and at Hull in the evening.

***Journal*, 12 Sep 1784:** I hastened to Kingswood, and preached under the shade of that double row of trees which I planted forty years ago. How little did any one of them think that they would answer such an intention! The sun shone as hot as it used to even in Georgia; but his rays could not pierce our canopy. Our Lord, meantime, shone upon many souls, and refreshed them that were weary.

The hand of God?

On several occasions during the middle period of his ministerial travels Wesley attributed changes in the weather, particularly relief from heat brought by clouds, to the influence of God on the elements. To what extent he honestly believed that God would intervene in such a specific personal way is left to the reader to decide.

***Journal*, 24 Apr 1755:** We rode in less than four hours the eight miles (so called) to Newell-Hay. Just as I began to preach, the sun broke out, and shone exceeding hot on the side of my head. I found, if it continued, I should not be able to speak long, and lifted up my heart to God. In a minute or two it was covered with clouds, which continued till the service was over. Let anyone who please call it *chance:* I call it an answer to prayer.

***Journal*, 16 Jun 1757:** I found myself much out of order, till the flux stopped at once without any medicine. But still being weak, and the sun shining extremely hot, I was afraid I should not be able to go round by Kelso. Vain fear! God took care for this also. The wind, which had been full east for several days, turned this morning full west; and blew just in our face: And about ten the clouds rose, and kept us cool till we came to Kelso.

***Journal*, 2 Jul 1759:** I rode to Durham, and went at one to the meadow by the river side, where I preached two years ago. The congregation was now larger by one half; but the sun was so scorching hot on my head, that I was scarce able to speak. I paused a little, and desired God would provide us a covering, if it was for his glory. In a moment it was done; a cloud covered the sun, which troubled us no more. Ought voluntary humility to conceal these palpable proofs, that God still heareth the prayer?

***Journal*, 8 Aug 1760:** [*At Kinsale*] About nine sharp, a storm having put an end to their exercise, I went to the soldiers in the field. I stood so near the intrenchments of the fort, that they could hear within as well as without. The sun indeed shone extremely hot on my head; but presently a cloud interposed. And when I began to be chill (for the wind was high and sharp) it removed till I wanted it again. How easily we may see the hand of God in small things as well as great! And why should a little pointless raillery make us ashamed to acknowledge it?

Wet and dry weather

In the same way that John Wesley was affected by heat and cold he was also often hindered by wet conditions that made travelling and preaching difficult.

However, in his early years he seemed to be resilient to the rigours of American conditions:

***Journal*, 2 Aug 1736:** In walking to Thunderbolt I was in so heavy a shower, that all my clothes were as wet as if I had gone through the river. On which occasion I cannot but observe that vulgar error, concerning the hurtfulness of the rains and dews of America. I have been thoroughly wet with these rains more than once; yet without any harm at all. And I have lain many nights in the open air, and received all the dews that fell; and so, I believe, might any one, if his constitution was not impaired by the softness of a genteel education.

Of course, rainy conditions became a common source of comment once he returned to England.

Letter, Bristol, 9 Apr 1739: … And the day being very cold and stormy (besides that much rain had fallen in the night) many who designed it were hindered from going to Hanham Mount, which is at least four miles distant from the town.

Letter, Bristol, 30 Apr 1739: The rain on Tuesday morning made them not expect me at Bath; so that we had not above a thousand or twelve hundred in the meadow …

Letter, Bristol, 28 May 1739: The rain prevented many from coming to Baptist Mills; but twelve or fifteen hundred stayed … His sending forth lightning with the rain did not hinder about fifteen hundred poor sinners from staying with me at Rose Green.

Letter, Bristol, 4 Jun 1739: The rain was so violent on Saturday that our congregation in the Bowling Green consisted of only nine hundred or a thousand.

Letter, Bristol, 2 Jul 1739: I preached the first time under the sycamore-tree near the school at Kingswood, during a violent rain, on those words of Isaiah, ‘As the rain cometh down from heaven, and returneth not thither, but watereth the earth and maketh it bring forth and bud ...'

By the 1740s his constitution was becoming more susceptible to wet weather. There are many instances of him describing himself as becoming completely soaking wet but he rarely seemed to be any the worse for it. He obviously did not wear any special waterproof clothing even though a type of oilskin was becoming available. There is only one rather grudging reference, in January 1764, to his use of an oilskin 'sou'wester' while suffering with a painful face.

***Journal*, 15 Apr 1744:** At five I preached at Gwennap, on a little hill, near the usual place. It rained from the time I began till I concluded. I felt no pain while I spoke, but the instant I had done, and all the time I was with the society, my teeth and head ached so violently, that I had hardly any senses.

***Journal*, 23 Mar 1750:** We took horse at five. It rained incessantly all the way we rode. And when we came on the great mountain, four miles from the town, (by which time I was wet from my neck to my waist,) it was with great difficulty I could avoid being borne over my mare's head, the wind being ready to carry us all away; Nevertheless, about ten we came safe to Dannabull, praising him who saves both man and beast.

***Journal*, 7-8 Sep 1751:** I rode in a stormy afternoon to St Just. But the rain would not let me preach abroad, either that evening or on *Sunday* morning ... I began at Newlyn about five. About the middle of the sermon there was a vehement shower of rain and hail; But the bulk of the congregation stood quite still, every man in his place.

***Journal*, 26-28 Sep 1752:** In the evening I called sinners to repentance in the main street at Bandon. On *Thursday* and *Friday* the rain drove us into the market-house. Indeed, I hardly remember two dry days together since I landed in Ireland.

***Journal*, 19 May 1753:** I preached at Pocklington again, and rode on to Whitgift-Ferry. It rained a great part of the way; and just as we got to the water, a furious shower began, which continued above half an hour … We set out from Whitgift soon after four; but the violent rain which attended us till after seven made the road so dirty and slippery, that our horses could hardly keep their feet; so that it was nine before we reached Epworth.

***Journal*, 11 Aug 1753:** The rain stopped at twelve, and gave me an opportunity of preaching in the market place at Camelford. I only saw one person in the congregation who was not deeply serious. That one (which I was sorry to hear) was the Curate of the Parish. Almost as soon as we set out, we were met by such a shower of rain as I never saw before in Europe. But it did us no hurt: We came very well, though very wet, to St Gennis.

***Journal*, 28 Aug 1758:** I scarce saw such a rain in Europe as we had for a considerable part of this morning. In one of the main streets the water ran with a stream capable of turning a mill. However, having appointed to preach at noon, in Newton, about six miles from Swansea, I was determined not to break my word, though I supposed but a few would attend: But I was mistaken.

***Journal*, 30 Jun 1760:** I have rarely seen so heavy rain in Europe as we had in the way to Tubbercurraugh. I was quickly wet to my toes' end; but the day clearing, I was dry again in a few hours.

***Journal*, 26 Aug 1763:** We designed to take horse at four, but the rain poured down, so that one could scarce look out. About six however, we set out and rode through heavy rain to St Clare … But when we had rode three or four miles, I recollected that I had heard of a ford, which would save us some miles riding. We inquired of an old man, who soon mounted his horse, showed us the way, and rode through the river before us.

***Journal*, 13 Jan 1764:** [*at Dorking*] I went at noon into the street … At first two or three little children were the whole of my congregation: But it quickly increased, though the air was sharp, and the ground exceeding wet: and all behaved well but three or four grumbling men, who stood far off that they disturbed none but themselves.

I had purposed to preach there again in the morning; but a violent storm made it impracticable. So, after preaching at Mitcham in the way, I rode back to London.

It is interesting to note that John Wesley judged the fitness of his companions by an 'equivalent age' compared to himself, so a person younger than him could be classified as older if he felt he was less fit for travel. For someone with such a keen interest in geography it is surprising that his estimates of distances sometimes proved to be way off the mark:

***Journal*, 16 Feb 1769:** Supposing we had but five-and-forty miles to Yarmouth, I did not set out till near seven: But it proved threescore; likewise it rained all day, and part of the road was very bad. However, God strengthened both man and beast: So we reached it before six in the evening. As we were both thoroughly wet, I was a little afraid for my companion, who was much older than me, though he had not lived so many years. But neither of us was any worse.

***Journal*, 1 May 1770:** I rode on to Aberdeen, and spent the rest of the week there. It fell out well, for the weather was uncommon: We had storms of snow or rain every day. And it seems the weather was the same as far as London. So general a storm has scarce been in the memory of man.

***Journal*, 8 Nov 1770:** I set out for London. It rained almost all the day; and in the afternoon so impetuously, (the rain being driven by a furious wind,) that it was with difficulty that we reached Lakenheath.

As Wesley kept such a complete record of events, including the weather, his comments on spells of continuous similar weather conditions can probably be taken as being fairly accurate:

***Journal*, 1 Jun 1771:** This is the twelfth day we have had continued rain, together with March winds. I dined at Rabin, near Castlebar, one of the pleasantest seats in Connaught.

***Journal*, 23-24 Jun 1778:** I went on to Tanderagee, one of the pleasantest towns in Ireland … For exactly two months, we have had only two days without rain.

***Journal*, 28 Aug 1781:** Between nine and ten we had such a storm of rain, as I do not remember to have seen in Europe before. It seemed ready to beat in the windows of the chaise, and in three minutes drenched our horsemen from head to foot. We reached Truro, however, at the appointed time.

As was the case for relief of excessively hot weather Wesley frequently attributed lack of rain at an appropriate time to 'divine intervention':

***Journal*, 17 Feb 1741:** In the afternoon I reached Oxford, and leaving my horse there, set out on foot for Stanton-Harcourt. The night overtook me in about an hour, accompanied with heavy rain. Being wet and weary, and not well knowing my way, I could not help saying in my heart, (though ashamed of my want of resignation to God's will,) O that thou wouldest 'stay the bottles of heaven'; or, at least give me light, or an honest guide, or some help in a manner thou knowest! Presently the rain ceased; the moon broke out, and a friendly man overtook me, who set me upon his own horse, and walked by my side till we came to Mr Gambold's door.

***Journal*, 2 Nov 1745:** (*Near Newcastle*) The rain was stayed to-day also, from nine to ten, (it fell both before and after) while I preached.

***Journal*, 28 Sep 1748:** I took horse between three and four, and came to St Mewan at eight. It rained all the time I was walking to the green, which was the usual place for preaching. But the moment I began to speak the rain ceased, and it did not begin again till I had done speaking. It rained with little intermission all the day after; which made the roads so bad that it was pretty dark when we came within two miles of Crimble-Passage.

***Journal*, 2 Sep 1750:** I rode to Tiverton … The meadow was then full from side to side, and many stood in the gardens and orchards round. It rained in the day several times; but not a drop fell while I was preaching. Here is an open door indeed! May no man be able to shut it!

***Journal*, 20 Jun 1761:** This evening it also rained at Hutton-Rudby, till seven, the hour of preaching: but God heard the prayer; and from the time I began we only had some scattering drops.

***Journal*, 8 Jun 1776:** As we rode to Sheep-Hill, we saw and heard at a distance, much thunder, and rain, and lightning. The rain was before and behind, and on each side: But none fell upon us. About six I preached at Sheep-Hill. It rained hard very near us; but not a drop came upon us. After eight I reached Newcastle, thoroughly tired.

***Journal*, 1 Jul 1777:** I preached in the evening at Marton, near Buckingham; the thunder, attended with heavy rain, was likely to rob us of our whole congregation. We cried to God. The thunder and rain ceased, and we had a fair sunshiny evening.

***Journal*, 26 Jun 1789:** Between Newtown and Moyle we had such rain as I never knew in Europe before. In two minutes the horsemen were drenched from head to foot. We dined at the Rector's, Dr Wilson, a man of very uncommon learning. At six he took me in his coach to the Castle-yard, where a numerous congregation soon assembled. Almost as soon as I began to preach, the rain began. Observing the people begin to scatter, I prayed aloud, that God would 'stay the bottles of heaven.' He did so: The people returned; and we had a comfortable, refreshing shower of heart-reviving love.

Making the best of it

On some, not that infrequent, occasions the rain continued throughout Wesley's outdoor services. In these cases he sometimes tried to put a brave face on it or even turn the weather to his apparent advantage:

***Journal*, 26 Jun 1739:** I preached near the house we had a few days before began to build for a school, in the middle of Kingswood, under a little sycamore tree, during a violent storm of rain, on those words, 'as the rain cometh down from heaven, and returneth not thither, but watereth the earth and maketh it bring forth and bud. So shall my word be that goeth out of my mouth: It shall not return to me void. But it shall accomplish that which I please, and it shall prosper in the thing whereto I sent it.'

***Journal*, 19 Aug 1769:** About eight I preached at Neath; about three in the church at Bridge-End; (where the rain doubled the congregation by stopping the harvest-work).

***Journal*, 6 May 1783:** [*Near Dublin*] The unusually large congregation in the evening was plentifully watered with the dew of heaven. I found a particular concern for the children, many of whom willingly attended.

***Journal*, 26 May 1789:** The rain prevented my preaching abroad at Brooksborough, although the congregation was exceeding large. Part of them were sheltered by a spacious turf-house, and the rest little regarded the rain; for the Lord watered us, in an uncommon degree, with the dew of his blessing.

Thunderstorms

Wesley obviously experienced very severe thunderstorms when in Georgia, and often compared ones he experienced later with them. Occasionally he saw advantages in them in that they frightened the people into attending service:

***Journal*, 19 Jul 1762:** [*At Tullamore*] Between two and three in the morning was such a thunder and lightning as I never knew in Europe. The crack and the flash were the same instant. Most of the houses shook; and yet no hurt was done in the whole town: But some good was done; for at five o'clock the preaching-house was quite filled; and the inward voice of the Lord was mighty in operation. This also was 'a glorious voice'.

The following unusual extracts appear to include a direct observation of some form of non-fatal lightning strike/near miss, and three possible descriptions of local 'ball lightning' or other similar effects:

***Journal*, 29 Jan 1736:** [*At the end of a 'hurricane' in the Atlantic*]

Toward the end of it, we had that appearance on each of the masts, which (it is thought) the ancients called Castor and Pollux. It was a small ball of white fire, like a star. The mariners say, it appears either in a storm, (and then commonly upon the deck,) or just at the end of it; and then it is usually on the masts or sails.

***Journal*, 23 Jul 1777:** I preached about one at Margam. In the road between this and Bridge-End, we had the heaviest rain I ever remember to have seen in Europe: And it saved John Prickard's life; for presently man and beast were covered with a sheet of lightning: but as he was thoroughly wet, it did him no harm.

***Journal*, 14 Apr 1782:** [*At St Helen's*] I was waked at two in the morning, as was Mr Broadbent also, by a very loud noise, like a vast crack of thunder, accompanied by a flash of bright light. It made the whole room shake, and all the tables and chairs therein jar. But (what is strange) none in the house, or in the town, heard it besides us.

***Journal*, 6 Mar 1788:** About the middle of the discourse, while there was on every side attention still as night, a vehement noise arose, none could tell why, and shot like lightning through the whole congregation. The terror and confusion were inexpressible. You might have imagined it was a city taken by storm … In about six minutes the storm ceased, almost as suddenly as it rose, and, all being calm, I went on without interruption.

There are many general reports of thunderstorms throughout the journal. For example these two separated by some thirty-five years:

***Journal*, 17 Nov 1755:** As we were walking towards Wapping, the rain poured down with such violence, that we were obliged to take shelter till it abated. We then held on to Gravel-Lane; in many parts of which the waters were like a river. However we got on pretty well, till the rain put out the candle in our lantern. We then were obliged to wade through all, till we came to the chapel-yard. Just as we entered it, a little streak of lightning appeared in the south-west. There was likewise a small clap of thunder, and a vehement burst of rain, which rushed so plentifully through our shattered tiles, that the very vestry was all in a float. Soon after I began reading Prayers, the lightning flamed all round it, and the thunder rolled over our heads. When it grew louder and louder, perceiving many strangers to be much affrighted, I broke off the prayers after the Collect, 'Lighten our darkness, we beseech thee , O Lord'; and began applying, 'The Lord sitteth above the water-flood; the Lord remaineth a king for ever.' Presently the lightning, thunder, and rain ceased, and we had a remarkably calm evening. It was observed, that exactly at this hour, they were acting *Macbeth* in Drury-Lane; and just as the mock-thunder began, the Lord began to thunder out of heaven. For a while it put them to a stand; But they soon took courage, and went on.

***Journal*, 22 Jun 1790:** I crossed over to Scarborough. The congregation in the evening was unusually small, being not yet recovered from the blessed fruits of the election. This was the hottest day we have had this year: And about one in the afternoon, the thunder, which had long lain at a distance, came near, with thick flashes of lightning, and impetuous rain: The thunder continued in one roll for an hour and a quarter. I never heard the like before, since my return from America.

Wesley showed a real interest in all types of natural phenomena including the weather, to the extent that he went out of his way to obtain reports and visit locations of extreme events to include in his writings:

***Journal*, 3 Mar 1748:** [*Part of the story of Mrs Holloway, near Holyhead*] One day she was sitting in her house, a flash of lightning killed a cat which sat just by her, and struck her to the earth, scorching her flesh in many parts, and yet not at all singeing her clothes. When she came to herself, she could not but acknowledge the loud call of God: But her seriousness did not continue long; her acquaintances soon laughed her out of it.

The following extract is a fine example of Wesley going further than just reporting an exceptional meteorological event but going on to analyse the data and come to a useful conclusion:

> ***Journal*, 30 Sep 1762:** About midnight I was waked by loud thunder, which continues about a quarter of an hour at Tiverton. In other places, we were afterwards informed, it continued a great part of the night. Yet comparing various accounts, I found the main shock was at the same time for near a hundred miles. So it seems there was a train of clouds for at least that space, which, by an electrical touch, were all discharged at once.

> ***Journal*, 14 Jul 1783:** I took a little journey into Oxfordshire, and found the good effects of the late storms. The thunder had been uncommonly dreadful; and the lightning had tore up a field near High Wycombe, and turned the potatoes to ashes.

and continuing:

> ***Journal*, 16 Jul 1783:** I went on to Witney. There were uncommon thunder and lightning here last Thursday (10th); but nothing to that which were there on Friday night (11th). About ten the storm was just over the town; and both the bursts of thunder and lightning, or rather sheets of flame, were without intermission. Those that were asleep in the town were waked, and many thought the day of judgment was come. Men, women and children, flocked out of their houses, and kneeled down together in the streets. With the flames, the grace of God came down also

in a manner never known before; and the impression was general, so it was lasting: It did not pass away with the storm; but the spirit of seriousness, with the grace of supplication, continued. A prayer meeting being appointed for the *Saturday* evening, the people flocked together; so that the preaching-house was more than filled ... When I came on the *Wednesday*, the same seriousness remained on the generality of the people.

Letter, Bristol, 20 Jul 1783: I do not remember any storm which travelled so far as that on the 10th. It has been in almost all parts of England, but especially at Witney, near Oxford. The next night they had a far greater, which seemed to cover the whole town for four hours with almost one uninterrupted blaze; and it has made such an impression on high and low, rich and poor, as had not been known in the memory of man.

Floods

The frequent result of heavy rain was flooding which, with poor roads, often made Wesley's journeys difficult and dangerous:

***Journal*, 18 Feb 1746:** We pushed on through thick and thin, and with much difficulty got to Stanley. Thence, after an hour's stop, we hastened on. The brooks were so swoln [*sic*] with the late rains, that the common roads were impassable; but our guide, knowing the country, carried us round about through the fields, so that we escaped the dangerous waters, and soon after sunset came (wet and dirty enough) to Evesham.

***Journal*, 30 Sep 1748:** [*Report of floods on 23 Jul 1747 – not for readers of a sensitive disposition*]:
We rode in the afternoon to Woodley. We saw by the way many marks of the late flood; of which John Bennet, who was then upon the place, gave us the following account:

On Saturday, the 23 July last, there fell for about three hours, in and about Hayfield, in Derbyshire, a very heavy rain, which caused such a flood as had not been seen by any now living in those parts.

'The rocks were loosened from the mountains: One field was covered with huge stones from side to side.

'Several water-mills were clean swept away, without leaving any remains.

'The trees were torn up by the roots, and whirled away like stubble.

'Two women of a loose character were swept away from their own door and drowned. One of them was found near the place; the other was carried seven or eight miles.

'Hayfield churchyard was all torn up, and the dead bodies swept out of their graves. Some were hanging on trees; others left in meadows or grounds; some partly eaten by dogs or wanting one or more members.'

***Journal* 14, Jun 1751:** The waters were greatly out in the road, so that the York coach was overturned just before us; the bridge it should have gone over being under water: Yet no passenger was hurt, only dropping wet, being all thrown into the river. We were to pass the same river a few miles off; and which way to do it we knew not. But just as we came to the place we overtook two gentlemen who had hired a guide. So we followed them as close as we could, and crossed it without difficulty.

***Journal*, 23 Aug 1756:** We rode forward to Redditch. It had rained all the way, so that Mr Walsh was obliged to go to bed as soon as we came in. Having dried some of our clothes, Mr Bruce and I took horse again about two; having one with us who knew the by-roads, the common road being impassable through the floods. About five we came to a broad water, which our guide did not care to pass. Mr Bruce, seeing a foot-bridge, walked over it, leading his horse by a long rein through the water: But in an instant the horse disappeared. However, he soon emerged and gained the bank. I rode through, at a small distance, very safely; and in the evening preached at Evesham.

A difficult river crossing

On several occasions flooding made river crossings quite dangerous:

***Journal*, 13 Apr 1759:** We took boat; but could not cross over. The violence of the stream, swollen by the late rains, bore us down in spite of all we could do. Having striven against it a considerable time, we were obliged to cast anchor. After waiting some time, we got near the shore, and were towed up to the place of landing. A toilsome day was followed by a comfortable night. At half-hour after eight the House at Epworth was well filled.

***Journal*, 16 Jan 1764:** Between twelve and one we crossed the Eynsham-Ferrry. The water was like a sea on both sides. I asked the ferryman, 'Can we ride the causeway?' he said, 'Yes, Sir; if you keep in the middle.' But this was the difficulty, as the whole causeway was covered with water to a considerable depth. And this in many parts ran over the causeway with the swiftness and violence of a sluice. Once my mare lost both her fore feet, but she gave a spring, and recovered the causeway: Otherwise we must have taken a swim; for the water on either side was ten or twelve feet deep. However, after one or two plunges more, we got through and came safe to Witney.

The following extract gives a detailed account of the impact of, and a suggested reason for, the severe flooding in the North of England:

***Journal*, 7 Dec 1771:** I read today of a circumstantial account of the late inundations in the north of England, occasioned by the sudden and violent overflowing of the three rivers, the Tees, the Wear, and the Tyne. All these have their rise within a few miles of each other, in a mountain at the head of Teesdale and Weardale; on which there was nothing more than a little mizzling rain, till the very hour when the rivers rose, and poured down such an amazing quantity of water as utterly astonished the people of Sunderland, at the mouth of the Wear, overflowed all the lower part of Newcastle-upon-the-Tyne, and filled the main street of Yarm, upon the Tees, with water nine or ten feet deep. Such an overflowing of these rivers none ever saw before, nor have we an account of any such in history.

Rain was not the cause of this; for there was next to none at the head of the rivers. What was the cause we may learn from a letter wrote at this time by a Clergyman in Carlisle: 'Nothing is so surprising as what lately happened at Solway-Moss, about ten miles from Carlisle. About four hundred acres of this Moss [*peat bog*] arose to such a height above the adjacent level, that at last it rolled forward like a torrent, and continued in its course above a mile, sweeping along with it houses and trees, and everything in its way. It divided itself into islands of different extent, from one to ten feet in thickness. It is remarkable, that no river or brook runs either through or near the Moss.'

To what cause then can any thinking man impute this, but to an earthquake? And the same doubtless it was, which, about the same time, wrought in the bowels of that great mountain, whence those rivers rise, and discharged from thence that astonishing amount of water.

***Journal*, 25 Nov 1774:** I set out between eight and nine in a one-horse chaise, the wind being high and cold enough. Much snow lay on the ground, and much fell as we crept along over the fen banks.

Honest, Mr Tubbs wouldst needs walk and lead the horse through the water and mud up to his mid-leg, smiling and saying, 'We fen-men do not mind a little dirt.' When we had gone about four miles, the road would not admit of a chaise. So I borrowed a horse and rode forward; but not far, for all the grounds were under water. Here therefore I procured a boat full twice as large as a kneading-trough. I was at one end, and a boy at the other, who paddled me safe to Erith.

Letter, London, 9 Feb 1789: It was providence indeed that the flood did not begin in the night rather than in the day. So it is that judgement is usually mixed with mercy.

Dry weather

Being very aware of the outdoor environment and the conditions of the countryside, Wesley, on several occasions, commented on exceptional dry spells:

***Journal*, 1 Aug 1764:** It was with difficulty I reached Cowbridge about one, where the congregation was waiting. I found they had

had heavy rain great part of the day; but very little fell upon us. Nor do I remember that from the beginning of March to now we have been in more than one heavy shower either in England, Scotland, or Wales.

***Journal*, 23 Jul 1765:** [*In Donnard, Co. Wicklow*] About the middle of the sermon, the rain (which was exceedingly wanted) began, and drove us into our friend's house.

***Journal*, 27 May 1783:** I preached at Loughborough in the morning, and at Mountsorrel at one. While I was preaching, the rain, which was so wanted, began, and continued for eight-and-forty hours.

Fogs

Another aspect of water encountered by Wesley was fog, which could cause significant travelling problems, sometimes involving divine intervention:

***Journal*, 27 Sep 1749:** I took horse at half past three. There was no moon, or stars, but a thick mist; so that I could see neither road, nor anything else; but I went as right as if it had been noon-day. When I drew nigh Penruddock-Moor, the mist vanished, the stars appeared and the morning dawned; so I imagined the danger was past; but when I was on the middle of the moor, the mist fell again on every side, and I quickly lost my way. I lifted up my heart. Immediately it cleared up and I soon recovered the high-road.

***Journal*, 29 Sep 1749:** I kept on as I could till I came to the brow of Hatside. So thick a fog then fell, that I was quickly out of all road, but knew not which way to turn. But I knew where help was to be found, in either great difficulties or small. The fog vanished in a moment and I saw Gamblesby at a distance.

***Journal*, 10 Nov 1755:** I preached at Wells ...The frost was very severe, accompanied with such a fog as perhaps the oldest man there never saw before. The lamps could not be seen across the street, and hardly the ground by those who had lights in their hands. Many lost their way, when they were just at their own doors. And it was almost as hard to breathe as to see. How easy it is for God to punish a sinful nation, even without employing an arm of flesh!

***Journal*, 27-30 Mar 1760:** I rode to Liverpool in order to embark for Dublin. We were desired to be on board by nine on *Saturday* morning: But the wind falling, and a fog coming on, we gained a little more time; so we had one more solemn opportunity in the evening.

~ *Sunday*, 30: The fog was gone, and the wind fair.

***Journal*, 22 Oct 1770:** I took horse a little before five, in an exceeding thick fog; but it was gone by noon. The rain, which was suspended all day, began again when we came to Whittlebury.

Wind and calm

Wind on land

Wesley was frequently subjected to windy weather while on his travels around the countryside. He occasionally took considerable risks in continuing with journeys on horseback in gale conditions, when a less adventurous traveller would have given up, in order to make his next engagement on time:

***Journal*, 5 Feb 1742:** I set out, and with some difficulty reached Chippenham on Saturday evening; the weather being extremely rough and boisterous, that I had much ado to sit on my horse.

***Journal*, 27 Dec 1742:** I rode to Horsley. The house being too small, I was obliged again to preach in the open air; but so furious a storm have I seldom known. The wind drove upon us like a torrent; coming by turns from east, west, north, and south; the straw and thatch flew round our heads; so that one would have imagined it could not be long before the house must follow; but scarce any one stirred, much less went away, till I dismissed them with the peace of God.

***Journal*, 29 Dec 1742:** I took horse for Tanfield. More than once I was only [*just*] not blown off my horse. However at three I reached Leigh, and explained to a multitude of people the salvation which is through faith. Afterwards I met the society in a large upper room, which rocked to and fro with the violence of the storm. But all was calm within.

***Journal*, 17 Mar 1752:** The rain continued without intermission, till we came to Enstone. Soon after we set out from thence, it was succeeded by so vehement a wind, as on Broadway-Hill often drove us clear out of the path, and was ready to carry away both horse and rider. But our strength was as our day; and before six in the evening we came unhurt to Evesham.

***Journal*, 30 Nov-1 Dec 1768:** I rode to Dover, and came in just before a violent storm began … The storm was ready to bear away both man and beast. But it abated about noon; so that, after preaching at Margate, I had a pleasant ride to Canterbury.

***Journal*, 18 Apr, 1774:** The minister of Heptonstall sent word that I was welcome to preach in his church. It was with difficulty we got up the steep mountain; and when we were upon it the wind was ready to bear us away. The church was filled, not with curious but serious hearers. No others would face so furious a storm.

***Journal*, 31 Dec 1778:** We concluded the old year with a solemn watch-night, and began the new with praise and thanksgiving. We had a violent storm that night. The roaring of the wind was like loud thunder. It kept me awake half an hour; I then slept in peace.

As on other occasions, Wesley occasionally attributed changes in the weather to God's influence:

***Journal*, 6 Oct 1788:** I preached at Loddon, North-Cove, and Lowestoft. When I came to the town, it blew a storm; and many cried out, 'So it always does when he comes.' But it fell suddenly as it rose; for God heard the prayer.

Being interested in the weather in general Wesley sometimes found space to include reports of past stormy conditions in his *Journal* and letters:

***Journal*, 21-24 Apr 1757:** I rode to Liverpool ...The upper part of the high spire of the church was blown down in the late storm. The stones, being bound together by strong iron cramps, hung waving in the air for some time. Then they broke through roof, gallery, pews, and pavement, and made a deep dint in the ground.

Letter, London, 11 Nov 1775: There was something very awful in that storm which I suppose reached all over England and Ireland; although it seems not to have been altogether so violent in London as in most other places.

To end this section describing windy events on land there is an entry in 1757 describing more unusual conditions:

Wind squalls causing dust storms as a precursor to a storm

> ***Journal*, 19 Jun 1757:** But here we had a new kind of inconvenience: Every four or five minutes a strong wind covered us over with a shower of dust; so that it was not easy to look up, or keep one's eyes open. But not long after the rain began, which constrained me to preach within, at Newcastle.

Wind at sea

The eighteenth century was a time of much use of sea passages for local as well as international travel. Journey times were totally dependent on the wind strength and direction, or, of course, the lack of any wind. It was still a risky business at times, even around British shores let alone in the open ocean.

Gales and storms

Wesley was indeed fully aware throughout his voyaging days of the risks involved and knew of ships that had foundered:

> ***Journal*, 20 Nov 1735:** We fell down into Yarmouth road, but the next day were forced back to Cowes. During our stay here there were several storms; in one of which two ships in the Yarmouth road were lost.

A little more detail is provided in a letter written on board his ship:

> **Letter, on board *Simmonds*, 20 Jan 1736:** Whilst we were in Cowes road, there were several storms, in one of which two ships were cast away on the back of the island, as we should have been if it had not pleased God to detain us in that safe station.

***Journal*, 22 Oct 1743:** The same day [*as John himself had a dangerous ferry crossing*], and, as near as we could judge, the same hour, the boat in which my brother was crossing the Severn, at the New-Passage, was carried away by the wind, and in the utmost danger of splitting on the rocks. But the same God, when all human hope was past, delivered them as well as us.

***Journal*, 26 Oct 1743:** She [*Mrs S.*] set out in a ship, bound for Hull. A storm met them by the way; the ship sprung a leak; but though it was near the shore, on which so many people flocked together, yet the sea ran so exceedingly high, that it was impossible to make any help. Mrs S. was seen standing on the deck, as the ship gradually sunk; and afterwards hanging by her hands on the ropes, till the masts likewise disappeared. Even then, for some moments, they could observe her floating on the waves, till her clothes, which buoyed her up, being thoroughly wet, she sunk – I trust, into the ocean of God's mercy.

***Journal*, 10 Oct 1752:** We had another violent storm: It made Mr P.'s house rock to and fro, though it was a new, strong house, and covered on all sides with hills, as well as with trees. We afterwards heard, that several ships were lost on the coast. Only one got into the harbour, but grievously shattered, her rigging torn in pieces, and her main mast gone by the board.

Letter, London, 12 Oct 1787: A second storm drove us to the Isle of Purbeck, just where the Indiaman was lost. There I had the opportunity of preaching to a little Society (at Swanage), which I had not seen for thirteen years.

We now turn to the most exciting accounts of Wesley's adventures, those during the many stormy sea crossings he endured. Firstly, and especially, his descriptions of the severe Atlantic storms on his way to and from America. There were five storms on the outward passage, the second of which could easily have resulted in the end of the John Wesley ministry before it had really even begun. It is interesting to note how quickly Wesley had picked up the sailing jargon for incorporation in his descriptions.

He was obviously frightened about the prospect of his transatlantic crossing:

> ***Journal*, 23 Nov 1735:** At night I was awaked by the tossing of the ship and the roaring of the wind, and plainly showed I was unfit, for I was unwilling, to die.

> ***Journal*, 17 Jan 1736:** Many people were very impatient at the contrary wind. At seven in the evening they were quieted by a storm. About nine the sea broke over us from stem to stern; burst through the windows of the state cabin, where three or four of us were, and covered us all over, though a bureau sheltered me from the main shock. About eleven I lay down in the great cabin, and in a short time fell asleep, though very uncertain whether I should wake alive, and much ashamed of my unwillingness to die. Oh how pure in heart must he be, who would rejoice to appear before God at a moment's warning! Toward morning, 'he rebuked the winds and the sea, and there was a great calm.'

> ***Journal*, 23 Jan 1736:** In the evening another storm began. In the morning it increased, so that they were forced to let the ship drive. I could not but say to myself, 'How is it thou hast no faith?' being still unwilling to die. About one in the afternoon, almost as soon as I had stepped out of the great cabin door, the sea did not break as usual, but came with a full, smooth tide over the side of the ship. I was vaulted over with water in a moment, and so stunned that I scarce expected to lift up my head again, till the sea should give up her dead. But thanks be to God, I received no hurt at all. About midnight the storm ceased.

Storm at sea

***Journal*, 25 Jan 1736:** At noon our third storm began. At four it was more violent than before … The winds roared round about us, and (what I never heard before) whistled as distinctly as if it had been a human voice. The ship not only rocked to and fro with utmost violence, but shook and jarred with so unequal, grating a motion, that one could not but with great difficulty keep one's hold on anything, nor stand a moment without it. Every ten minutes came a shock against the stern or side of the ship, which one would think should dash the planks to pieces. At this time a child, privately baptized before, was brought to be received into the church … the sea broke over, split the main-sail in pieces, covered the ship, and poured between the decks, as if the great deep had already swallowed us up.

During this storm Wesley visited, and was impressed by the calmness of, the German passengers in comparison with the panic of the British.

***Journal*, 29 Jan 1736:** About seven in the evening, we fell in with the skirts of a hurricane. The rain as well as the wind was extremely violent. The sky was dark in a moment, that the sailors could not so much as see the ropes, or set about furling the sails. The ship must, in all probability, have overset, had not the wind fell suddenly as it arose.

***Journal*, 30 Jan 1736:** We had another storm, which did us no harm than splitting the fore-sail. Our bed being wet, I laid me down on the floor, and slept sound till morning.

Finally:

***Journal*, 5 Feb 1736:** Between two and three in the afternoon, God brought us all safe into the Savannah river. We cast anchor near Tybee Island, where the groves of pines, running along the shore made an agreeable prospect.

Wesley had some significant brushes with dangerous windy conditions whilst in America on some of his journeys around the coastline. For instance:

***Journal*, 22 Apr 1736:** About four in the afternoon we entered upon Doboy Sound. The wind, which was right ahead of us, was so high, when we were in the middle of it, and the sea so rough, being in at the inlet, that the boat was on the point of sinking every moment. But it pleased God to bring us safe to the other side in half an hour, and to Frederica the next morning.

***Journal*, 26 Jul 1736:** My brother and I set out for Charlestown, in order [*sic*] to his embarking for England; but the wind being contrary, we did not reach Port-Royal, forty miles from Savannah, till Wednesday evening. The next morning we left it. But the wind was so high in the afternoon, as we were crossing the neck of St Helena's sound, that our oldest sailor cried out, 'Now every one must take care of himself.' I told him, 'God would take care of us all.' Almost as soon as the words were spoken, the mast fell. I kept on the edge of the boat, to be clear of her when she sunk, (which we expected every moment,) though with little prospect of swimming ashore, against such a wind and sea. But, 'How is it thou hadst no faith?' The moment the mast fell, two men caught it, and pulled it into the boat; the other three rowed with all their might, and 'God gave command to the wind and seas'; so that within an hour we were safe on land.

On his journey home to England Wesley again hit very high winds which this time he called 'a proper hurricane'. To be able to survive actual hurricane-force winds would be open to doubt, but if they can be taken at face value, the changes in wind direction he reports on certainly point at least to being beneath an extremely deep and quickly-travelling depression:

***Journal*, 13 Jan 1738:** We had a thorough storm which obliged us to shut all close; the sea breaking over the ship continually. I was at first afraid; but cried out to God, and was strengthened. Before ten, I lay down: I bless God without fear. About midnight we were awakened by a confused noise of seas and wind and men's voices, the like of which I had never heard before. The sound of the sea breaking over and against the sides of the ship, I could compare to nothing but large cannon, or American thunder. The rebounding, starting, quivering motion of the ship much resembled what is said of earthquakes. The captain was upon deck in an instant. But his men could not hear what he said. It blew a proper hurricane; which beginning at south-west, then went west, north-west, north, and, in a quarter of an hour, round by the east to the south-west point again. At the same time the sea running (as they term it) mountain high, and that from many different points at once, the ship would not obey the helm;

nor indeed could the steersman, through the violent rain, see the compass. So he was forced to let her run before the wind, and in half an hour the stress of the storm was over.

His transatlantic adventure finally ends with the following account which gives an idea of the problems of relying on sail-power:

***Journal*, 29 Feb 1738:** We saw the English land once more; which, about noon, appeared to be the Lizard-point. We ran by it with a fair wind; and at noon, the next day, we made the west end of the Isle of Wight.

Here the wind turned against us, and in the evening blew fresh, so that we expected (the tide being likewise strong against us) to be driven some leagues back in the night: but in the morning to our great surprise, we saw Beachy-head just before us, and found we had gone forwards near forty miles.

Toward evening was a calm; but in the night a strong north wind brought us safe to the Downs.

Throughout his ministry Wesley made many sea crossings and coastal journeys, especially to and from Ireland. He includes many accounts in his writings, a few of which are given below and paint a vivid picture of the stormy weather and complicated crossings he encountered:

***Journal*, 28 Mar-6 Apr 1750:** About eleven we were called to go on board, the wind being quite fair: And so it continued till we were just out of the harbour. It then turned west and blew a storm … We fought our way four or five leagues toward Ireland; but were driven back in the afternoon to the very mouth of the harbour. Nevertheless, the wind shifting one or two points, we ventured out again; and by midnight we were got about half seas over; but the wind then turning full against us, and blowing hard, we were driven back again, and were glad, about nine, to get into the bay once more. … In the night there was a vehement storm. Blessed be God we were safe on shore! …

~ **Fri. 6:** I preached near Llanerellymadd at noon, and at Trefollwin in the evening. Observing at night the wind was changed, I rode to Holyhead early in the morning. A ship was just ready to sail; so we went on board, and in the evening landed at Dublin.

***Journal*, 27-30 Apr 1756:** On our way to Holyhead, one met and informed us, the packet sailed the night before. I said, 'Perhaps it may carry me for all that.'… The packet did sail the night before and got more than half sea-over. But the wind turning against them, they were glad to get back this afternoon. I scarce remember so violent a storm as blew all night long. The wind continued contrary the next day … We left the harbour about twelve, having six or seven Officers and an abundance of passengers on board. The wind was full west, and there was great possibility of a stormy night. So it was judged best to put back; but one gentleman making a motion, to try a little longer, in a short time brought us all over to his opinion. So they agreed to go out, and 'look for a wind.' The wind continued westerly all night. Nevertheless, in the morning we were within two leagues of Ireland! Between nine and ten I landed at Howth.

Letter, Bangor Ferry, 15 Oct 1777: As I believe both you and our friends in Dublin will be pleased to hear of our safe landing, I will first give you a short account of what has befallen us hitherto. We went on board the packet about seven. The wind was full in our teeth and blew high. However, tacking to and fro, by the help of the tide, we got out to sea before noon. When the tide turned against us, we made little way, but beat up and down till night. The wind rising higher and higher, and the sea in proportion, my fellow passengers were all sick: I, by the blessing of God, was not sick at all. However, as I could not read, I lay down about seven, but (which is very uncommon with *me*) could not sleep till past midnight.

In the morning, the wind being still very high and directly contrary, we found ourselves out of course, having been driven too far to the southward. But being under the Carnarvonshire shore, we were in tolerably smooth water. About noon, the wind coming a point to the south, we put out to sea again. But the storm increased, and about four carried away our bow-sprit and tore one of the sails all to bits. The captain and his men with all quietness and composure quickly repaired the loss. And before six, by the good providence of God, we landed safe at Holyhead.

***Journal*, 11 Jul 1787:** I went down with my friends, having taken the whole ship, and went on board the *Prince of Wales*, one of the Parkgate packets. At seven we sailed with a fair moderate wind. Between nine and ten I lay down, as usual, and slept till near four, when I was waked by an uncommon noise, and found the ship lay beating upon a large rock, about a league from Holyhead. The Captain, who had not long lain down, leaped up; and, running upon the deck, when he saw how the ship lay, cried out, 'Your lives may be saved, but I am undone!' Yet no sailor swore, and no woman cried out. We immediately went to prayer; and presently the ship, I know not how, shot off the rock, and pursued her way, without any more damage, than the wounding of a few of her outside planks. About three in the afternoon we came safe to Parkgate; and in the evening went on to Chester.

Smaller boats were used for short coastal journeys and estuary crossings which were more susceptible to strong tides or even moderately bad weather which could easily put the lives of travellers in danger, for example:

> ***Journal*, 6-7 Oct 1758:** I designed to go in a wherry to the Isle of Wight; but the watermen were so extravagant in their demands, that I changed my mind and went in the hoy: And it was well I did; for the sea was so high, it would not have been easy for a small boat to keep above water. We landed at two, and walked on, five little miles, to Newport …
>
> We hired a small fisher-boat at Wotton-Bridge, there being scarce any wind. But it increased more and more when we were on the sea, which was seven miles over. Our cock-boat danced on the waves, and must have sunk, if one large wave had come over her: but God suffered it not. We landed in two hours, and walked away to Gosport.

Calm conditions

A sailing ship cannot make any progress, and may indeed drift dangerously in any tidal flow, if there is not enough wind for it to make way. Wesley gives examples of this situation, sometimes 'solving' the difficulty through prayer:

> ***Journal*, 15 Apr 1749:** But no sooner were we out of harbour, than it turned south-west and blew a storm. Yet we made forward, and about one o'clock came within two or three leagues of land. The wind wholly failed; a calm suddenly following a storm, produced such a motion as I never felt before. But it was not long before the wind sprung up west, which obliged us to stand away from the Skerries. When we wanted a league of shore it fell calm again, so that there we rolled about till past sunset. But in the night we got back to Dublin Bay.

> ***Journal*, 9 Apr 1785:** Between nine and ten we went on board the *Claremont* packet: but it was a dead calm till past ten on *Sunday,* 10, when the company desired me to give them a sermon. After sermon I prayed that God would give us a full and speedy

passage. While I was speaking the wind sprung up; and in twelve hours brought us to Dublin bay.

***Journal*, 6 Sep 1787:** Went on board with a fair, moderate wind; but we had but just entered the ship when the wind died away. We cried to God for help, and it presently sprung up, exactly fair, and did not cease till it brought us to Penzance Bay.

4
Physical geographer

As a keen observer of all aspects of the environment he travelled through, Wesley included many interesting descriptions of the physical landscape.

Landscape change

Wesley made comments on the physical geography, both forming processes and appearance, of a range of landscapes that he encountered.

He was very moved by the grandeur and scale of Land's End, though his precise memory failed leading him to a vast overestimate of the amount of erosion that had occurred between his first and last visits:

> ***Journal*, 11 Sep 1743:** We went down, as far as we could go safely, toward the point of the rocks at the Land's-End. It was an awful sight! But how will these melt away, when God arises to Judgement! The sea between does indeed 'boil like a pot.' 'One would think the deep to be hoary.' But 'though they swell, yet can they not prevail. He hath set their bounds, which they cannot pass.'

> ***Journal*, 10 Sep 1757:** We rode to the Land's End. I know of no natural curiosity like this. The vast ragged stones rise on every side, when you are near the point of land, with green turf between, as level and smooth as if it were the effect of art. And the rocks which terminate the land are so torn by the sea, that they appear like great heaps of ruins.

A difficult climb

***Journal*, 25 Aug 1785:** About nine I preached at Mousehole, where there is now one of the liveliest societies in Cornwall. Hence we went on to the Land's End, in order to [*see*] which we clambered down the rocks to the very edge of the water. And I cannot think but that the sea has gained some hundred yards since I was here forty years ago.

Also:

***Journal*, 28 Jan 1756:** I preached at noon at Dover, to a very serious but small congregation. We afterwards walked up to the castle on top of a mountain. It is an amazingly fine situation; and from hence we had a clear view of that vast piece of the cliff, which a few days ago divided from the rest and fell down upon the beach.

Dramatic real or postulated changes in the fortunes of coastal areas due to sand encroachment were noted on more than one occasion:

> ***Journal*, 21 Nov 1748:** I set out for Leigh, in Essex ... Here was once a deep open harbour; but the sands have long since blocked it up, and reduced a once flourishing town to a small ruinous village.

> ***Journal*, 14 Jul 1777:** In riding along the side of Newport-Bay, I observed on the ground a large quantity of turfs. These are found by removing the sand above the high-water mark, under which there is a continued bed of turf, with roots of trees, leaves, nuts, and various kinds of vegetables. So that it is plain the sea is an intruder here, and now covers what was once dry land. Such probably was the whole Bay a few centuries ago. Nay, it is not at all improbable, that formerly it was dry land from Aberystwith to St. David's Point.

> ***Journal*, 18 Jun 1780:** About noon I preached in the market-place at Wainfleet. Once a large sea-port town, till the harbour was blocked up by sand. We now passed into marshland, a fruitful and pleasant part of the county.

Earthquakes and landslips

Some of the most dramatic of geological events are earthquakes and Wesley, despite very limited understanding of their processes, took a keen interest in them. He was greatly affected by what he considered to be their theological implications and consequences. He was keen to hear from those who had experienced the events to check on the extent of the quake and consistency of their reports, and, if possible, travel to see the sites for himself:

> ***Journal*, 8 Feb 1750:** It was about a quarter after twelve, that the

earthquake began at the skirts of the town. It began in the south-east, went through Southwark, under the river, and then from one end of London to the other. It was observed at Westminster and Grosvenor-Square a quarter before one. (Perhaps, if we allow for the differences in the clocks, about a quarter of an hour after it began in Southwark.) There were three distinct shakes, or wavings to and fro, attended with a hoarse, rumbling noise like thunder. How gently does God deal with this nation! O that our repentance may prevent heavier marks of his displeasure!

***Journal*, 8 Mar 1750:** To-day God gave the people of London a second warning; of which my brother wrote as follows:

'This morning, a quarter after five, we had another shock of an earthquake, far more violent than that of February 8. I was just repeating my text, when it shook the Foundery so violently, that we all expected it to fall upon our heads. A great cry followed from the women and the children. I immediately cried out, "Therefore we will not fear, though the earth be moved, and the hills be carried into the midst of the sea: for the Lord of hosts is with us, the God of Jacob is our refuge." He filled my heart with faith, and my mouth with words, shaking their souls as well as their bodies.'

The earth moved westward, then east, then westward again, through all London and Westminster. It was a strong and jarring motion, attended with a rumbling noise, like that of distant thunder. Many houses were much shaken, and some chimneys thrown down, but without any farther hurt.

There is one highly detailed account of a British earthquake in the *Journal* which has subsequently been extensively studied with modern observations confirming the accuracy of his report. Whether the event was indeed of seismic origin or a major landslip is still in doubt. It caused controversy at the time when Wesley's published account in *The Gentleman's Magazine* was questioned by another correspondent. This elicited a virulent repost from Wesley 'I aver to be punctually true, having been an eye-witness of every particular of it. And if F.D. [*the offending correspondent*] will set his name, and aver the contrary, I will make him ashamed, unless shame and he have shook hands and parted.'

[*A fuller account can be found in Roger Cooper's* 'John Wesley at Whitestone Cliff, North Yorkshire, 1755', *published by University of York, Borthwick Paper No. 91, 1997.*]

Although rather extensive, the majority of Wesley's account deserves to be reproduced in full as it demonstrates so well the precise physical observations and descriptions he was able to produce:

> ***Journal*, 2 Jun 1755:** On Tuesday, March 25th last, being the week before Easter, many persons observed a great noise near a ridge of mountains in Yorkshire, called Black-Hamilton. It was observed chiefly in the south-west side of the mountain, about a mile from the course where the Hamilton races are run; near a ridge of rocks, commonly called Whitston-Cliffs, or Whitston-White-Mare; two miles from Sutton, about five from Thirsk.
>
> The same noise was heard on Wednesday, by all who went that way. On Thursday, about seven in the morning, Edward Abbot, weaver, and Adam Bosomworth, bleacher, both of Sutton, riding under Whitston-Cliffs, heard a roaring, (so they termed it,) like many cannons, or loud and rolling thunder. It seemed to come from the cliffs; looking up to which, they saw a large body of stone, four or five yards broad, split and fly off from the very top of the rocks. They thought it strange, but rode on. Between ten and eleven a larger piece of rock, about fifteen yards thick, thirty high, and between sixty and seventy broad, was torn off, and thrown into the valley.
>
> About seven in the evening, one who was riding by observed the ground to shake exceedingly; and soon after several large stones or rocks, of some tons weight each, rose out of the ground.

Others were thrown on one side, others turned upside down, and many rolled over and over. Being a little surprised, and not very curious, he hastened on his way.

On Friday and Saturday the ground continued to shake, and the rocks to roll over one another. The earth also clave asunder in very many places, and continued to do so until Sunday morning.

Being at Osmotherly, seven miles from the cliffs, on Monday, June 2, and finding Edward Abbot there, I desired him, the next morning, to show me the way thither. I walked, crept, and climbed, round and over a great part of the ruins. I could not perceive, by any sign, that there was ever any cavity in the rock at all; but one part of the solid stone is cleft from the rest, in a perpendicular line, and smooth, as if cut with instruments: Nor is it barely thrown down, but split into many hundred pieces; some of which lie four or five hundred yards from the main rock.

The ground nearest the cliff is not raised, but sunk considerably beneath the level: But at some distance it is raised in a ridge of eight or ten yards high, twelve or fifteen broad, and near a hundred long. Adjoining to this lies an oval piece of ground, thirty or forty yards in diameter, which has been removed, whole as it is, from beneath the cliff, without the least fissure, with all its load of rocks; some of which were as large as the hull of a small ship. At a little distance is a second piece of ground, forty or fifty yards across, which has also been transplanted entire, with rocks of various sizes upon it, and a tree growing out of one of them. By the removal of one or both of these, I suppose the hollow near the cliff was made.

All round them lay stones and rocks, great and small; some on the surface of the earth, some half sunk into it, some almost covered, in a variety of positions. Between these the ground was cleft asunder in a thousand places: Some of the apertures were nearly closed again; some gaping as at first. Between thirty and forty acres of land, as is commonly supposed, (though some reckon above sixty) are in this condition.

On the skirts of these, I observed, in abundance of places, the green turf (for it was pasture land) as it were pared off two or three inches thick, and wrapped round like sheets of lead. A little farther, it was not cleft or broken at all, but raised in ridges five or six foot long, exactly resembling the graves in a church-yard. Of these there is a vast number.

That part of the cliff from which the rest is torn, lies so high and is now so bright a colour, that it is plainly visible to all the country round, even at the distance of several miles. We saw it distinctly not only from the street in Thirsk, but for five or six miles as we rode towards York. So we did likewise in the Great North-Road, between Sandhutton and Northallerton.

Without any knowledge of earthquake mechanisms Wesley then goes on to consider processes involving fire, water, or air as the natural cause of the event but could not reach a definite solution. He consequently concludes with the following explanation:

What then could be the cause? What, indeed, but God who arose to shake terribly the earth; who purposely chose such a place, where there is so great a concourse of Nobility and Gentry every year; and wrought in such a manner, that many might see it and fear; that all who travel one of the most frequented roads in England, might see it, almost whether they would or no, for many miles together. It must likewise for many years, maugre all the art of man, be a visible monument of His power; all that ground being now so encumbered with rocks and stones, that it cannot be either ploughed or grazed. Nor will it serve any use, but to tell all that see it, Who can stand before this great God?

***Journal*, 18 May 1757:** I rode in the afternoon to Halifax, over the huge, but extremely pleasant and fruitful, mountains to Heptonstall ... Here I was informed of the earthquake the day before. – On Tuesday, May 17, many persons in several parts, within five or six miles, heard a strange noise under the ground, which some compared to thunder, others to the rumbling of carts. Quickly after, they had felt the earth rock under them, and wave to and fro. Many who were within doors heard their pewter and glass clatter; many in the fields felt the ground shake under their feet; and all agreed as to the time, though they knew nothing of each other's account.

19 May: [*At Haslingden*] Here I learned that the earthquake observed near Heptonstall had been sensibly felt by very many persons from Bingley, three miles eastward of Kieghley [*sic*] to the neighbourhood of Preston. It was every where preceded by a hoarse rumbling, about three o'clock; so that in a few minutes it had run east to west, between fifty to sixty miles.

***Journal*, 3 Sep 1757:** [*At St Agnes*] Some who live here gave me an account of the earthquake on July 15. There was first a rumbling noise under the ground, hoarser and deeper than common thunder. Then followed a trembling of the earth, which afterward waved once or twice to and fro so violently that one said he was obliged to take a back-step, or he should have fallen down; and another, that the wall against which he was leaning seemed to be shrinking from him.

Early in 1768 witnesses of the catastrophic great earthquake of Lisbon had reached Britain:

***Journal*, 8 Feb 1768:** He was at Lisbon during the great earthquake, just then sitting in his night-gown and slippers. Before he could dress himself, part of the house he was in fell, and blocked him up. By this means his life was saved; for all who had run out were dashed to pieces by the falling houses.

In common with many in England, he was greatly affected by the

Lisbon disaster in which tens of thousands were probably killed. In his subsequent sermon 'On the cause and cure of earthquakes', Wesley returns to the theme in his 1755 entry blaming the wrath of God induced by the sin of the populace on the events.

This last earthquake extract graphically describes both its effects on the landscape and their dramatic hydrological consequences:

> ***Journal*, 9-10 Jul 1773:** I went, by moderate stages, from Liverpool to Madeley; where I arrived on Friday 9. The next morning we went to see the effects of the late earthquake: Such it undoubtedly was. On Monday, 27, at four in the morning, a rumbling noise was heard, accompanied with sudden gusts of wind, and wavings of the ground. Presently the earthquake followed, which only shook the farmer's house, and removed it entire about a yard; but carried the barn about fifteen yards, and then swallowed it up in a vast chasm; tore the ground into numberless chasms, large and small; in the large, threw up mounts, fifteen or twenty feet high; carried an hedge, with two oaks, above forty feet, and left them in their natural position. It then moved under the bed of the river; which making more resistance, received a ruder shock, being shattered in pieces and heaved up about thirty feet from its foundations. By throwing this, and many oaks, into its channel, the Severn was quite stopped up, and constrained to flow backward, till, with incredible fury, it wrought itself a new channel. Such a scene of desolation I never saw. Will none tremble when God thus terribly shakes the earth?

As this last extract shows, as late in his life as his last year of preaching travels, Wesley was still a keen observer of landscape change – in this case what he considered an advantageous one!

> ***Journal*, 18 Jun 1790:** I preached at Stokesley in the morning; and then went on to Whitby. It was very providential, that part of the adjoining mountain fell down, and demolished our old

preaching-house, with many houses besides; by which means we have one of the most beautiful chapels in Great Britain, finely situated on the steep side of the mountain.

General landscape and geology

Throughout his travelling life Wesley had a keen interest in detailing the physical attributes of the landscape, either seen first hand or read about in books – sometimes also using them to illustrate aspects of the power of God.

Descriptions of such places – sometime showing his lack of knowledge of true comparative heights of various land forms – include:

***Journal*, 10 Dec 1735:** We sailed from Cowes, and in the afternoon passed the Needles. Here the ragged rocks, with the waves dashing and foaming at the foot of them, and the white side of the island rising to such a height, perpendicular from the beach, gave a strong ideas of 'Him that spanneth the heavens and holdeth the waters in the hollow of his hand!'

***Journal*, 15 Sep 1755:** [*At Cubert, Devon*] We walked an hour near the sea-shore, among those amazing caverns, which are full as surprising as Poole's Hole, or any other in the Peak of Derbyshire. Some part of the rock in these natural vaults glitters as bright and ruddy as gold: part is a fine sky-blue; part green; part enamelled, exactly like mother-of-pearl; and a great part, especially near the Holy Well, (which bubbles up on the top of a rock, and is famous for curing either scorbutic or scrofulous disorders,) is crusted over, wherever the water runs, with an hard, white coat, like alabaster.

***Journal*, 12 Jul 1762:** I went to Dunmore-Cave, three or four miles from Kilkenny. It is full as remarkable as Poole's Hole, or any other in the Peak. The opening is round, parallel to the horizon, and seventy or eighty yards across, In the midst of this, there is a kind of arch, twenty or thirty feet high. by this you enter into the first cave, nearly round and forty or fifty feet in diameter. It is encompassed with spar stones, just like those on the sides of Poole's Hole. On one side of the cave is a narrow passage, which goes under the rock two or three hundred yards; on the other side, an hollow, which no one has ever been able to find an end of. I suppose this hole too, as well as many others, was formed by the waters of the deluge, retreating into the great abyss, with which probably it communicates.

Arthur's Seat, Edinburgh

***Journal*, 16 Jun 1766:** I took a view of one of the greatest natural curiosities in the kingdom; what is called Arthur's seat; a small, rocky eminence, six or seven yards across, on the top of an exceeding high mountain, not far from Edinburgh. The prospect from the top of the castle is large, but it is nothing in comparison with this.

***Journal*, 14 May 1770:** After ten years' enquiry, I have learned what are the Highlands of Scotland. Some told me, 'the Highlands begin where you cross the Tay'; others, 'when you cross the North Esk'; and others 'when you cross the river Spey': But all of them missed the mark. For the truth of the matter is, the Highlands are bounded by no river at all, but by Carns, or heaps of stones, laid in a row, south-west and north-east, from sea to sea. These formerly divided the kingdoms of the Picts from that of the Caledonians, which included all the country north of the Carns; several whereof are still remaining. It takes in Argyleshire, most of Perthshire, Murrayshire, with all the north-west counties. This is called the Highlands because a considerable part of it (though not the whole) is mountainous. But it is not more mountainous than North-Wales, nor than many parts of England and Ireland: Nor do I believe it has any mountain higher than Snowdon hill, or the Skiddaw in Cumberland. Talking Erse [*Gaelic*], therefore, is not the thing that distinguishes these from the Lowlands. Neither is this or that river; both the Tay, the Esk, and the Spey running through the Highlands, not south of them.

***Journal*, 6 Jun 1778:** I was desired to take a ride to the celebrated Giant's Causeway. It lies eleven English miles from Coleraine. When we came to the edge of the precipice, three or four poor boys were ready to hold our horses, and show us the way down. It being dead low water, we could go anywhere, and see everything to the best advantage. It is doubtless the effect of subterraneous fire. This manifestly appears from many of the stones which composed the pillars that are now fallen down: these evidently bear the mark of fire, being burnt black on one or other surface. It appears likewise from the numerous pumice-stones scattered among the pillars: Just such pillars and pumices are found in every country which is, or ever was, subject to volcanoes.

***Journal*, 12 May 1779:** After preaching at Cuthburton and in Teesdale, I went a little out of my way, to see one of the wonders of nature. The river Tees rushes down between two rocks, and falls sixty feet perpendicular into a basin of water, sixty feet deep.

Wesley was a voracious reader of all kinds of books, whether when he was indoors or on the move. These sometimes included topics within physical geography, and he occasionally included comments on what he had read within the *Journal* which demonstrate the very limited real knowledge of many subjects that existed in his time. For instance:

***Journal*, 28 Jul 1756:** I read Mr Barton's ingenious *Lectures on Lough-Neagh*, near Lurgan, which turns wood into stone, and cures the King's evil, and most cutaneous distempers. Under part of this lake first there is a stratum of firm clay, and under that stratum of trees four foot thick, all compacted into one mass, doubtless by the pressure of the incumbent earth, (perhaps water too) which it has probably sustained ever since the General Deluge.

***Journal*, 17 May 1768:** I looked over Dr Shaw's Travels: Great part of them is very dull and unentertaining; but some remarks are extremely curious. I was a little surprised at one of them; namely, that the celebrated Mount Atlas is not higher than many of our English mountains, and nothing near so high as the Alps. But it was much farther from Rome: So travellers might make it as high as the moon; and few in Italy would contradict them.

***Journal*, 17 Jan 1770:** In a little journey, which I took into Bedfordshire, I finished Dr Burnet's *Theory of the Earth*. He is doubtless one of the first-rate writers, both as to sense and style;

his language is remarkably clear, unaffected, nervous and elegant. And as to his theory, none can deny that it is ingenious, and consistent with itself. And it is highly probable, 1. That the earth arose out of the chaos in some such manner as he describes: 2. That the antediluvian earth was without high or abrupt mountains, and without sea, being one uniform crust enclosed by the great abyss: 3. That the flood was caused by the breaking of this crust, and its sinking into the abyss of waters: And, 4. That the present state of the earth, both internal and external, shows it to be the ruin of the former earth. This is the substance of his two former books, and this far I can go with him.

Hydrology and climatology

There are several occasions described in his *Journal* where Wesley appears to show some understanding of water-related (hydrological) matters:

***Journal*, 3 Dec 1737:** [*In America*] It now grew toward sunset; so we sat down, faint and weary, having had no food all day, except a gingerbread cake, which I had taken in my pocket. A third of this we had divided among us at noon; another third we took now; the rest we reserved for the morning; but we had met with no water all the day. Thrusting a stick into the ground, and finding the end of it moist, two of our company fell a digging with their hands, and, at about three feet depth, found water. We thanked God, drank, and were refreshed.

***Journal*, 26-28 Dec 17:** We went on with a small, fair wind, till Thursday in the afternoon; and then sounding, found a whitish sand at seventy-five fathom: But having had no observation for several days, the Captain began to be uneasy, fearing we might either get unawares into the Bristol Channel, or strike in the night on the rocks of Scilly.

~ **Sat 28:** Was another cloudy day; but about ten in the morning (the wind continuing southerly) the clouds began to fly just contrary to the wind, and, to the surprise of us all, sunk down under the sun, so that at noon we had an exact observation; and by this we found we were as well as we could desire, about eleven leagues south of Scilly.

***Journal*, 17 May 1756:** Walking up at the Red-House Walk, (which runs between two rows of meadows, with the river winding through them, and a chain of fruitful hills on the right hand and on the left,) I saw the plain reason why strangers usually complain of the unwholesomeness of the water in Cork. Many women were filling vessels with river-water (which is that commonly used in the city for tea and most other purposes) when the tide was at the height. Now, although, this is not salt, yet it cannot but affect both the stomach and bowels of tender persons.

***Journal*, 7 Jul 1757:** This day, between Helmsley and Kirkby-Moorside, we rode over a little river, which suddenly disappears; and, after running a mile under ground, rises again and pursues his course.

There are also occasional references to large or small scale climatological information which Wesley had obtained locally or from his reading:

***Journal*, 2 Dec 1737:** During this time I had frequent opportunities of making many observations and inquiries concerning the real state of this province, (which has been so variously represented,) the English settlements therein, and the Indians that have intercourse [*conversation*] with them. These I minuted down from time to time; a small extract of which I have subjoined.

1. Georgia lies in the 30th and 31st degree of north latitude. The air is generally clear, the rains being much shorter, as well

as heavier, than in England. The dews are great. Thunder and lightning are expected almost every day in May, June, July, and August. They are very terrible, especially to a stranger. During those months, from ten in the morning to four in the afternoon, the sun is extremely scorching. But the sea-breeze generally blows from ten till three or four. The winter is nearly of the same length as in England. But the mid-day sun is always warm, even when the mornings and evenings are very sharp, and the nights piercing cold.

***Journal*, 12 Jun 1764:** We rode through the pleasant and fertile county of Murray to Elgin. I never suspected before that there was any such country as this near and hundred and fifty miles beyond Edinburgh; a country which is supposed to have generally six weeks more sunshine in a year than any part of Great Britain.

***Journal*, 28 Apr 1772:** We walked through the Duke of Athol's gardens, in which was one thing I never saw before – a summer-house in the middle of a green-house, by means of which one might in the depth of winter enjoy the warmth of May, and sit surrounded with greens and flowers on every side.

***Journal*, 6 Jul 1781:** To-day I finished the second volume of Dr Robertson's *History of America*. … If it was worth while to mention a little thing, after things of so much greater importance, I would add, I was surprised that so sensible a writer, in enumerating so many reasons, why it is so much colder in the southern hemisphere than it is in the northern; why it is colder, for instance, at forty degrees south than at fifty north latitude; should forget the main, the primary reason, namely, the greater distance from the sun! For is it not well known, that the sun (to speak with the vulgar) is longer on the north side the line than the south? That he is longer in the six northern signs than the southern, so there is a difference (says Gravesande) of nine days? Now, if the northern hemisphere be obverted to the sun longer than the southern, does not this necessarily imply that the northern hemisphere will be warmer than the southern? And is not this the primary reason for being so?

Biogeography related topics

Throughout his life Wesley showed an interest in natural history, often noting details of plants or animals and the environments where they were found.

The awareness of the seasonality of natural events (phenology) is of considerable interest for climate change studies. Although Wesley does not formally record such events there are a few occasions where he does note seasonal information:

Letters, 12 Dec 1724: We still have very warm weather at Oxford; and a gentleman, now in the room with me, says that several of the flowers in his father's garden, who lives in town, are blown as if it were spring.

***Journal*, 2 Feb 1736:** Between two and three in the afternoon, God brought us all safe into the Savannah river. We cast anchor near Tybee Island, where the groves of pines, running along the shore, made an agreeable prospect, showing, as it were, the bloom of spring in the depth of winter.

***Journal*, 22 Oct 1748:** I spent an hour observing the various works of God in the Physic Garden at Chelsea. It would be a noble improvement of the design, if some able and industrious persons were into the full and accurate enquiry into the uses and virtues of all these plants: Without this, what end does the heaping them thus together answer, but the gratifying of an idle curiosity?

***Journal*, 26 Jul 1770:** I rode to Castle-Donnington; but hay-making had emptied the town, till a violent shower brought all the hay-makers home, who received the good word with gladness.

***Journal*, 12 Mar 1784:** Being at Samuel Rayner's, in Bradford, I was convinced of two vulgar errors; the one, that nightingales will not live in cages; the other that they only sing a month or two in the year. He has now three nightingales in cages; and they sing almost all day long, from November to August.

***Journal*, 11 May 1784:** I breakfasted at the first house I was invited to at Inverness ... In the afternoon we took a walk over the bridge into one of the pleasantest countries I have seen. It runs along by the side of the clear river, and is well cultivated and well wooded. Here we first heard abundance of birds welcoming the return of spring.

The *Journal* also contains brief descriptions of unusual plants that caught Wesley's interest:

***Journal*, 11 Jun 1764:** [*Near Nairn*] The house is an old castle, which stands on a little hill, with a delightful prospect all four ways; and the hospitable master has left nothing undone to make it still more agreeable. He showed us all his improvements, which are very considerable in every branch of husbandry. In his gardens many things were more forward than in Aberdeen, yea, or Newcastle. And how is it, that none but one Highland gentleman has discovered that we have a tree in Britain, as easily raised as an ash, the wood of which is full as fine a red as mahogany? namely, the laburnum. I defy any mahogany to exceed the chairs which he has lately made of this.

***Journal*, 11 Oct 1773:** I took a little tour through Bedfordshire and Nottinghamshire. Between Northampton and Towcester we met with a great natural curiosity, the largest elm I ever saw; it was twenty-eight feet in circumference; six feet more than that which was some years ago in Magdalen-College walks at Oxford.

***Journal*, 22 Feb 1775:** I had the opportunity of seeing Mr Gordon's curious garden at Mile End [*near Hertford*], the like of which I suppose is hardly to be found in England, if in Europe. One thing in particular I learned here, the real nature of the tea-tree. I was informed, 1. That the Green and Bohea are quite different species. 2. That the Bohea is much tenderer than the Green. 3. That the Green is an evergreen; and bears, not only in the open air, but in the frost, perfectly well. 4. That the herb of Paraguay likewise bears the frost, and is a species of tea. 5. And I observed that they are all species of bay or laurel. The leaf of Green tea is both the colour, shape, and size of a bay leaf: That of Bohea is smaller, softer, and of a darker colour. So is the herb of Paraguay, which is of a dirty green; and no larger than our common red sage.

***Journal*, 9 Jun 1775:** I preached at eight to a numerous congregation, in the market place at Dungannon; at eleven, and at five in the afternoon in the main street at Charlemount. I lodged at a gentleman's who showed me a flower, which he called a Gummy Cystus. It blooms in the morning, with a large beautiful, snow-white flower; but every flower dies in the evening. new flowers blow and fall every day. Does not the short-lived flower answer to that short-lived animal the Ephemeron-fly?

***Journal*, 24 Jul 1780:** [*At Bristol*] In the evening I saw one of the greatest curiosities in the vegetable creation – the Nightly Cereus. About four in the afternoon, the dry stem began to swell; about six, it gradually opened; and about eight, it was in its full glory. I think the inner part of this flower, which was snow-white, was about five inches in diameter; the yellow rays which surrounded it, I judged, were in diameter nine or ten inches. About twelve it began to droop, being covered with a cold sweat; at four it died away.

Wesley also occasionally made comments on birds and the environment he observed them in, sometimes going to quite great lengths to reach the necessary location:

***Journal*, 21 May 1772:** I went to the Bass, seven miles from it [*Dunbar*], which, in the horrid reign of Charles the Second, was the prison of those venerable men who suffered the loss of all things for a good conscience. It is a high rock surrounded by the sea, two or three miles in circumference, and about two miles from the shore. The strong east wind made the water so rough, that the boat could hardly live: And when we came to the only landing-place, (the other sides being quite perpendicular,) it was with much difficulty that we got up, climbing on our hands and knees. The castle, as one may judge by what remains, was utterly inaccessible. The walls of the chapel, and the Governor's house, are tolerably entire. The garden-walls are still seen near the top of the rock, with the well in the midst of it. And round the walls there are spots of grass, that feed eighteen or twenty sheep. But the proper natives of the island are Solund-geese [*gannets*], a bird about the size of a Muscovy-duck, which breeds by thousands, from generation to generation, on the sides of the rock. It is peculiar to these, that they lay but one egg, which they do not sit upon at all, but keep it under one foot, (as we saw with our eyes,) till it is hatched.

***Journal*, 24 Mar 1773:** A very remarkable paragraph was published in one of the Edinburgh Papers:

'We learn from the Rosses, in the county of Donegal, in Ireland, that a Danish man-of-war, called the *North Crown*, commanded

> by the Barron D'Ulfeld, arrived off those islands, from a voyage of discovery towards the Pole. They sailed from Bornholme, in Norway, the 1st of June, 1769, with stores for eighteen months, and some able astronomers, landscape-painters, and every apparatus suitable to the design; and steering N. by E. half E., for thirty seven days, with a fair wind and open sea, discovered a large rocky island, which having doubled, they proceeded W. N. W., till the 17th September, when they found themselves in a strong current, between two high lands, seemingly about ten leagues distant, which carried them at a prodigious rate for three days, when, to their great joy, they saw the main land of America, that lies between the most westerly part of the settlements on Hudson's River and California. Here they anchored, in a fine cove, and found abundance of wild deer and buffaloes, with which they victualled; and sailing southward, in three months got to the Pacific Ocean, and returned by the Straits of Le Maine and the West India Islands.
>
> They have brought many curiosities, particularly a prodigious bird, called a contor, or contose [*condor*], above six feet in height, of the eagle kind, whose wings, expanded, measure twenty-two feet four inches. After bartering some skins with the country people, for meal, rum, and other necessaries, they sailed for Bremen, to wait the thaw, previous to their return to Copenhagen.'

However, he was far more upset by and sceptical of other accounts of exploration as in this entry made only a few months later, on 17 Jan 1774, he recorded:

> Meeting with a celebrated book, a volume of Captain Cook's voyages, I sat down to read it with huge expectation. But how was I disappointed! I observed 1. Things absolutely incredible: 'A nation without any curiosity'; and, what is stranger still, (I fear related with no good design,) 'without any sense of shame! Men and women coupling together in the face of the sun, and in the

sight of scores of people! Men whose skin, cheeks and lips are white as milk.' Hume or Voltaire might believe this; but I cannot. I observed 2. Things absolutely impossible. To instance in one, for a specimen. A native of Otaheite is said to understand the language of an island eleven hundred degrees [*presumably miles*] distant from it in latitude; beside I know not how many hundreds in longitude! So that I cannot but rank this narrative with that of Robinson Crusoe; and account Tupis to be, in several respects, akin to his man Friday.

***Journal*, 30 Jun 1779:** Took a view of Flamborough-Head. It is an huge rock, rising perpendicular from the sea to an immense height, which gives shelter to an innumerable multitude of sea-fowl of various kinds.

***Journal*, 2 Jun 1781:** [*On the Isle of Man*] All the day I observed, wherever I was, one circumstance that surprised me: In England we generally hear the birds singing morning and evening; but here thrushes, and various other kinds of birds, were singing all day long. They did not intermit, even during the noon-day heat, where they had a few trees to shade them.

Wesley was also not averse to broadening his knowledge on natural history areas by visits to appropriate museum exhibits:

***Journal*, 9 Dec 1759:** The same day I spent part of the afternoon in the British Museum. There is a great library, a great number of curious manuscripts, many uncommon monuments of antiquity, and a whole collection of shells, butterflies, beetles, grasshoppers, &c., which the indefatigable Sir Hans Sloane, with such vast expense and labour, procured in a life of fourscore years.

***Journal*, 22 Jan 1780:** [*In London*] I spent an hour or two very agreeably in Sir Ashton Lever's museum. It does not equal the British Museum in size; nor is it constructed on so large a plan; as it contains no manuscripts, no books, no antiquities, nor any remarkable works of art. But I believe, for natural curiosities, it is not excelled by any museum in Europe; and all the beasts, birds, reptiles, and insects, are admirably well ranged and preserved: So that if you saw many of them elsewhere, you would imagine they were alive! The hippopotamus, in particular, looks as fierce as if he was just coming out of the river; and the old lion appears as formidable now as when he was stalking the tower.

Unusual environmental events

Wesley occasionally reported unusual extreme weather events he had become aware of. During one visit to Cornwall he collected two such reports, of tornado/water-spouts, which had caused considerable consternation and damage:

***Journal*, 15 Sep 1760:** I inquired concerning the uncommon storm, which was here on March 9, the last year. It began near the Land's End, between nine and ten at night, and went eastward not above a mile broad, over St Just, Morva, Zennor, St Ives, and Gwinear, whence it turned northward, over the sea. It uncovered all the houses in its way, and was accompanied with impetuous rain. About a mile south-east from St Ives, it tore up a rock, twelve or fourteen ton weight, from the top of a rising ground, and whirled it down upon another, which it split through, and at the same time dashed itself in pieces. It broke down the pinnacles of Gwinear church, which forced their way through the roof. And it was remarkable, the rain which attended it was as salt as any sea-water.

***Journal*, 17 Sep 1760:** At noon I preached on the cliff near Penzance, where no one now gives an uncivil word. Here I

> procured an account, from an eye witness, of what happened on the twenty-seventh of last month. A round pillar, narrowest at the bottom, of a whitish colour, rose out of the sea near Mousehole, and reached the clouds. One who was riding over the strand from Marazion to Penzance saw it stand for a short space, and then move swiftly toward her, till, the skirt of it touching her, the horse threw her and ran away. It had a strong sulphurous smell. It dragged with it abundance of sand and pebbles from the shore; and then went over the land, carrying with it corn, furze, or whatever it found in its way. It was doubtless a kind of water-spout; but a water-spout on land, I believe, is seldom seen.

In 1757 he reports on exceptional damage to vegetation done by 'burning winds' – presumably not only extremely strong, but exceptionally dry and perhaps salty:

> ***Journal*, 24-5 May 1757:** We rode from Keswick to Whitehaven … I was surprised to see, not only hedges and shrubs without a green leaf upon them, but abundance of trees likewise naked as in the depth of winter. Upon inquiring, I found that some time before, a violent wind had gone through all these parts, which not only threw down chimneys, walls, and barns, and tore up trees by the roots, but scorched everything green it touched as with fire, so that all the leaves immediately fell off, and not only bushes and fruit trees, but elms, oaks and firs, withered away to the very roots.

> ***Journal*, 7 and 15 Sep 1757:** [*Near Camborne*] I observed more and more the effects of that burning wind which was in these parts on Sunday, the 28th of last month. It not only scorched all the leaves of the trees, so as to bring mid-winter upon them in two hours, but burned up all the leaves of potatoes and cabbage, and every green thing which it touched! What a mercy that it did not come a month sooner! Then it would have left little work

for the reapers. As we rode toward Helstone, I think the sun was near as hot as it was at midsummer; yet all along the trees looked as in the depth of winter, that scorching wind having destroyed all it touched.

In the eighteenth century the aurora borealis (Northern Lights) tended to be thought of as exceptional meteorological phenomena by scientists and often as a portent of impending disaster or divine judgement by many of the general population. Wesley occasionally reported them among all his other records of unusual events:

***Journal*, 11 Oct 1752:** I rode to Cork once more, and was very fully employed all the day. The next morning we returned to Cove, and about noon got out of the harbour. We immediately found the effects of the late storm, the sea still boiling like a pot. The moon set about eight, but the Northern Lights abundantly supplied her place.

***Journal*, 24 Oct 1769:** [*At Northampton*] This evening there was such an Aurora Borealis as I never saw before: The colours, both the white, the flame-colour, and the scarlet, were so exceeding strong and beautiful. But they were awful too: So that abundance of people were frightened into many good resolutions.

5
Human geographer

John Wesley had a genuine and deep concern for the life and welfare of people from all backgrounds, which was demonstrated in his interest in many matters which would nowadays fall under the general term 'human geography'. The following sections give a picture of the spread of interests he had in this field of study.

In addition to the short *Journal* items meant for public study, and his letters intended for a small audience reproduced here, Wesley also published a range of tracts, some aimed at certain portions of the population, and also more general ones, including several on human geography topics:

Thoughts upon Liberty (1772), *Thoughts concerning the Origin of Power*, *Thoughts upon Slavery* (1774), *The Present State of Public Affairs* (1776), *Some Observations on Liberty* (1776) and *A serious Address to the People of England with regard to the State of the Nation* (1778).

He considered geography an important part of education and recommended *Randall's Geographical Grammar* [*Letters, 18 Feb 1756 and Jun 1764*] as a vital text book to be included in any course of study for both boys and girls.

There are also within his writings examples of fairly broad-brush slightly naive descriptions of the characters of those inhabiting various places. For instance the following *Journal* entry, describes those living in the area where the main author of this book lives!

Journal, 12 Oct 1774: The Isle (or properly Peninsula) of Purbeck is nine or ten miles broad, and perhaps twenty long, running nearly from north-east to south-west. Two mountains run almost the whole length, with valleys both between them and on each side, but poorly cultivated. The people in general are plain, artless, good-natured, and well-behaved. If the labourers here are zealous and active, they will surely have a plentiful harvest.

Population geography (demography)

Wesley developed a keen interest in estimations of the population of England, and potential errors in them, in addition to how the lives of the people themselves were changing:

***Journal*, 1 May 1776:** I set out early, and the next afternoon we reached Whitehaven; and my chaise-horses were no worse for travelling near a hundred and ten miles in two days.

In travelling through Berkshire, Oxfordshire, Bristol, Gloucestershire, Worcestershire, Warwickshire, Staffordshire, Cheshire, Lancashire, Yorkshire, Westmorland, and Cumberland, I diligently made two enquiries:

The first was, concerning the increase or decrease of the people; the second was, concerning the increase or decrease of trade.

As to the latter, it is, within these last two years, amazingly increased; in several branches in such manner as has not been known in the memory of man: such is the fruit of the entire civil and religious liberty which all England now enjoys!

And as to the former, not only in every city and large town, but in every village and hamlet, there is no decrease, but a very large and swift increase. One sign of this is the swarms of little children which we see in every place. Which, then, shall we most admire, the ignorance or the confidence of those that affirm population decreases in England? I doubt not but it increases full as fast here, as in any province in North America.

> ***Journal*, 9 Sep 1776:** I began, what I had long intended, visiting the society from house to house, setting apart at least two hours in a day for that purpose. I was surprised to find the simplicity with which one and all spoke, both of their temporal and spiritual state. Nor could I easily have known, by any other means, how great the work of God has wrought among them. I found exceeding little to reprove; but much to praise God for. And I observed one thing which I did not expect: In visiting all the families without Lawford-Gate, by far the poorest about the city, I did not find so much as one person who was out of work.
>
> Another circumstance I critically inquired into, What is the real number of the people? Dr Price says, (doubtless to encourage our good friends, the French and Spaniards,) 'The people of England are between four and five millions; supposing them to be four, or four and a half, on average in one house.' I found, in the families which I visited, about six in a house. But one who has lately made a more general inquiry, informs me, there are, without Lawford-Gate, seven in a house. The same information I received, from one who has lately made the inquiry, concerning the inhabitants of Redcliff. Now, if at four in a house, we are four millions, must we not, at seven in a house, be seven millions?
>
> But even if this is far short of the truth; for a plain reason, the houses are miscomputed. To give one instance: The houses without Lawford-Gate are computed to be a thousand. Now, at the sitting of the Justices, some years since, there were two hundred public-houses. Was then one house in five a public-house? No, surely; one in ten at the utmost. If so, there were two thousand houses; and, consequently, fourteen thousand persons. I believe, there are now full twenty thousand. And these are nothing near a quarter of the present inhabitants of Bristol.

Wesley's criticism of the published estimates of the population of England are quite specific and, even though it is based on a very limited sample of locations, he comes up with definite conclusions.

It is nowadays possible to go some way towards testing his hypotheses on population numbers. Published estimates for the population of England and Wales (not greatly above the numbers for England alone) are 6.2 million in 1771 rising to 7 million by 1781. This means that his estimate of 7 million for the population of England in 1776 was in actual fact not far off the mark. The first formal census occurred in 1801 giving a population of England of over 8 million.

It is also possible to look at Wesley's estimates of populations of English cities he mentions in his descriptions in the 1750s:

1. 'If the population of Liverpool continues to increase at the current rate [*in 1755*] in forty years it will equal that of Bristol.'

The estimated population of Bristol in 1750 is 45,000 which increases to 64,000 by 1801, whereas in contrast Liverpool only had 22,000 people in 1750 and this had already increased to 80,000 by 1801. Therefore Wesley's forecast for equality in numbers by 1795 must have easily been exceeded. By 1900 the population of Liverpool had risen to 700,000 which was over twice that of Bristol.

2. 'Portsmouth may probably contain half as many people as Bristol.'

Certainly by the 1801 census Portsmouth had a population of 35,000 which was well above half the Bristol population of 64,000.

It seems from these examples that he had a good appreciation of demographic issues and his comments on populations (though perhaps not on the vast numbers attending his sermons!) can be taken as worthy of taking note of.

The current populations of Bristol, Liverpool and Portsmouth are around 570,000, 1.2 million and 210,000, respectively, which compares with 8.3 million for London.

While on the topic of populations, it is of interest to record how the town of Savannah, where Wesley was based during his time in Georgia, has developed subsequently. The relatively small outpost grew to over 5,000 persons by 1800, 54,000 by 1900 and over 130,000 today.

Historical and economic geography

Wesley was always concerned about honesty among his clergy and quickly responded to any suggestion of wrong behaviour. The extract below discusses the crimes of poaching and also smuggling which was prevalent in Cornwall:

> **Letter, Tiverton, 21 Sep 1755:** By these dull wretches you have been an unspeakable loser. Perhaps it was in company of some of these that you first thought of taking a little sport, and catching a few fish, or killing a partridge or an hare. Miserable employment for a preacher of the gospel! For a Methodist preacher, above all others! Though I do not wonder if, after practising it for some time, you should be so infatuated as even to defend it. I am afraid these same poor creatures afterwards taught you (if that report be true) even to countenance that wickedness for which Cornwall stinks in the nostrils of all who fear God or love King George; I mean that of smuggling: though surely they could not persuade you to receive stolen goods!

Wesley's interests in historical geography extended to other parts:

> ***Journal*, 21 Apr 1760:** In riding to Rosmead, I read Sir John Davis's *Historical Relations concerning Ireland.* None who reads these can wonder, that, fruitful as it is, it was always so thinly inhabited; for he made it plain,

1. That murder was never capital among the native Irish; the murderer only paid a small fine to the Chief of his sept [*an Irish clan*].

2. When the English settled here, still the Irish had no benefit of English laws. They could not so much sue an Englishman. So the English beat, plundered, yea, murdered them, at pleasure.

3. Arose continual wars between them, for three hundred and fifty years together; and hereby both the English and Irish natives were kept few, as well as poor.

4. When they were multiplied during a peace of forty years, from 1600 to 1641, the general massacre, with the ensuing war, again thinned their numbers; not so few as a million of men, women, and children, being destroyed in four years' time.

5. Great numbers have ever since, year by year, left the land merely for want of employment.

6. The Gentry are continually driving away hundreds, yea, thousands, of them that remain, by throwing such quantities of arable land into pasture, which leaves them neither business or food.

This is that now dispeoples many parts of Ireland, of Connaught in particular, which, it is supposed, has scarce half the inhabitants as this day which it had fourscore years ago.

***Journal*, 24 Jul 1760:** I looked over that well-wrote book, Mr Smith's *State of the County and City of Waterford*. He plainly shows, that twelve hundred years ago Ireland was a flourishing kingdom. It seems to have been declining almost ever since; especially after it was torn into several independent kingdoms. Thenceforward it grew more and more wild and barbarous, for several hundred years. In Queen Elizabeth's time it began to revive; and it increased greatly both in trade and inhabitants, till the deadly blow which commenced on 23 October, 1641. Three hundred thousand protestants, by a moderate computation, were destroyed in less than a year; and more than twice as many Papists, within a few

> years following: Most of these were adults, and this was a loss which the nation has not recovered yet. Nay, it will probably require another century, to restore the number of inhabitants it had before.

The long war and poor harvests had caused much scarcity, and prices had risen to such an extent as to cause acute distress. This impelled him to write a lengthy letter to the editor of *Lloyd's Evening Post* and later to *The Leeds Mercury* on many aspects of the scarcity of provisions, discussing the reasons, as he saw them, for the high unemployment rates and high prices of bread, corn, oats, meat, eggs, and land. He also discussed the relation between high taxes and the national debt estimated to be four million pounds.

His proposed solutions were as follows:

Letter, Dover, 9 Dec 1772

1. Many thousand poor people are starving. Find them work and you will find them meat. They will then earn and eat their own bread.

2. But how shall their masters give them work without ruining themselves? Procure vent for it, and it will not hurt their masters to give them as much work as they can do; and this will be done by sinking the price of provisions, for then people will have money to buy other things too.

3. But how can the price of *wheat* be reduced? By prohibiting for ever that bane of health, that destroyer of strength, of life, and of virtue, *distilling*. Perhaps this alone will answer the whole design. If anything be more needful, may not all starch be made of rice, and the importation of this as well as of wheat be encouraged?

4. How can the price of *oats* be reduced? By reducing the number of horses. And may not this be effectually done (1) by laying a tax of ten pounds on every horse exported to France, (2) by laying an additional tax on gentleman's carriages. Not so

much for every wheel (barefaced shameless partiality!), but ten pounds yearly for every horse. And these two taxes alone would nearly supply as much as is now given for leave to poison His Majesty's liege subjects.

5. How can the price of *beef* and *mutton* be reduced? By increasing the breed of sheep and horned cattle. And this would be increased sevenfold if the price of horses was reduced, which it surely would be half in half by the method above mentioned.

6. How can the price of *pork* and *poultry* be reduced? First, by letting no farms above an hundred pounds a year. Secondly, by repressing luxury, either by example, by laws, or both.

7. How may the price of land be reduced? By all the methods named above, all of which tend to lessen the expense of house-keeping; but especially the last, restraining luxury, which is the grand source of poverty.

8. How may the taxes be reduced? By discharging half the national debt, and so saving at least two millions a year.

How can this be done, the wisdom of the great council of the land can best determine.

Around the 1770s Wesley made a point of looking on the positive side of the prevailing poor economic conditions, and wrote to contradict some of what he saw as the excesses of contemporary 'doom and gloom' commentaries:

***Journal*, 26-27 Oct 1772:** At twelve I set out in the stage coach, and in the evening came to Norwich. Finding abundance of people were out of work, and, consequently, in the utmost want, (such a general decay of trade having hardly been known in the memory of man,) I enforced, in the evening, 'Seek ye first the kingdom of God, and his righteousness; and these things shall be added unto you.' For many years I have not seen so large a congregation here, in the mornings as well as evenings. One reason of which may be this: Thousands of people, who, when they had fullness of bread, never considered whether they had any souls or not, now they are in want begin to think of God.

***Journal*, 2 Feb 1778:** This week I visited the society, and found a surprising difference in their worldly circumstances. Five or six years ago, one in three, among the lower ranks of people was out of employment; and the case was supposed to be nearly the same through all London and Westminster. I did not now, after all the tragic outcries of want of trade that fill the nation, find one in ten out of business; nay, scarce one in twenty, even in Spitalfields.

***Journal*, 17 Feb 1778:** I wrote *A Serious Address to the Inhabitants of England with Regard to the Present State of the Nation* – so strangely misrepresented both by ignorant and designing men – to remove, if possible, the apprehensions which have been so diligently spread, as if it were on the brink of ruin.

***Journal*, 28 Mar 1781:** I returned to Burslem. How is the whole face of this country changed in about twenty years! Since the potteries were introduced, inhabitants have continually flowed in from every side. Hence the wilderness is literally become a fruitful field. Houses, villages, towns have sprung up. And the country is not more improved than the people. The word of God has had free course among them. Sinners are daily awakened and converted to God; and believers grow in the knowledge of Christ.

Wesley was interested in all things economic and occasionally passed on what he had learned in his *Journal*. This extract describes the necessity of taking inflation into account when comparing past and present prices of goods. However, the use of the price of corn in an example was rather inappropriate as it is one of the commodities most prone to violent changes in price due to weather conditions:

Meeting and preaching to the populace

***Journal*, 2 Sep 1776:** In my way to Exeter, I read over an ingenious tract, containing some observations which I never saw before. In particular, that if corn sells for twice as much now as it did at the time of the Revolution, it is in effect no dearer than it was then, because we have now twice as much money; that if other things sell now for twice as much as they did then, corn ought to do so too; that though the price of all things increases as money increases, yet they are really no dearer than they were before; and, lastly, that to petition Parliament to alter these things, is to put them upon impossibilities, and can answer no end but that of inflaming the people against their Governors.

Closing this section with another report on the effect of the variable economic situation in English towns:

***Journal*, 20 Mar 1779:** We went to Birmingham, still increasing

> on every side. Hearing the cry of want of business, even in this as well as most other trading towns in England, I considered what the meaning of it should be; and the case seems plainly this: Two or three years ago, business poured into Birmingham, and consequently more hands were wanting; but when business returned into its usual channel, they were wanted no longer. These men therefore certainly wanted business, and spread the cry all over the town. The same must be the case at Manchester, Liverpool, and all other towns where there is an extraordinary trade for a time; It must subside again: and then arises the cry of want of business.

In a series of letters (*7 September, 25 September, 31 October 1789*) Wesley discusses what he sees as the ill effects of hops, and partly justifies the case for their use to be banned with the following economic argument:

> Does not the tax on hops bring in two or three hundred thousand pounds yearly into the Exchequer? Perhaps it does. And yet it may be not an advantage but a loss to the nation. So it certainly is if it breeds and increases grievous and mortal diseases, and thereby destroys every year thousands of His Majesty's liege subjects. May not gold be bought too dear? Are not one hundred thousand lives worth more than two hundred thousand pounds? Each of these men, had this poison been kept out of his reach, had he lived out all his days, would probably have paid more yearly in other taxes than he paid for leave to put himself out of the world.

Political geography

In addition to main addresses, and a very long letter *To a Friend*, [*Letter, December 1768*] on the topic of 'the present state of public affairs', Wesley occasionally made in his *Journal* general comments on the past and current political situation and the treatment and behaviour of the working classes:

***Journal*, 2 Oct 1754:** I walked to Old Sarum, which, in spite of common sense, without house or inhabitant, still sends two members to Parliament.

***Journal*, 25 Aug 1755:** I rode over the mountains, close by the sea to Looe, a town near half as large as Islington, which sends four Members to Parliament! And each county in North Wales sends one!

***Journal*, 15 Jun 1773:** I went to dreary Newtown. This place always makes me feel pensive. Even in Ireland I hardly see anywhere such heaps of ruins as here; they are considerably increased since I was here before. What a shadow is human greatness!

When I came to Belfast, I learned the real cause of the late insurrections in this neighbourhood. Lord Donegal, the proprietor of almost the whole county, came hither to give his tenants new leases. But when they came, they found two merchants of the town had taken their farms over their heads; so that multitudes of them, with their wives and children, were turned out into the wide world. It is no wonder that, as their lives were now bitter to them, they should fly out as they did. It is rather a wonder that they did not go much farther. And if they had, who would have been most at fault? Those who were without home, without money, without food for themselves and families? Or those who drove them to this extremity?

In the evening I preached to a numerous congregation in the new market-house, but trifling enough. Yet by degrees they sunk into seriousness. The greater part of them came again in the morning; and their behaviour was then remarkably decent.

***Journal*, 17 Aug 1776:** We found Mr Hoskins, at Cubert, alive;

> but just tottering over the grave. I preached in the evening, on 2 Cor. 5:1-4, probably the last sermon he will hear from me. I was afterwards inquiring, if that scandal of Cornwall, the plundering of wrecked vessels, still subsisted. He said, 'As much as ever; only Methodists will have nothing to do with it. But three months since a vessel was wrecked on the south coast, and the tinners presently seized all the goods; and even broke in pieces a new coach which was on board, and carried every scrap of it away.' But is there no way to prevent this shameful breach of all the laws both of religion and humanity? Indeed there is. The gentry of Cornwall may totally prevent it whenever they please. Let them only see that the laws be strictly executed upon the next plunderers; and after an example is made of ten of these, the next wreck will be unmolested. Nay, there is a milder way. Let them only agree together, to discharge any tinner or labourer that is concerned in the plundering of a wreck, and advertise his name, that no Cornish gentleman may employ him any more; and neither tinner or labourer will any more be concerned with that bad work.

Wesley wrote a lengthy letter to William Pitt, the First Lord of the Treasury, who had become prime minister in December 1783. This put forward in very definite terms his views and suggestions on economic matters such as land tax, the widespread avoidance of window and other taxes, the excess of distilling, and the danger of alcohol. He then goes on the discuss suicide and the fact that there is generally no ignominy associated with it, resulting in 'self murder' increasing daily. He suggests every 'self-murderer' should be ignominiously hanged in chains on public display to make others not take the same course. He includes, towards the end of his letter, the following comment:

> **Letter, Bath, 6 Sep, 1784:** Suppose your influence could prevent suicide by this means, and distilling by making it a felony,

you would do more service to your country than any Prime Minister has done these hundred years.

General human geography

Wesley always showed a real concern for those at the very bottom of the social scale, including prisoners of war and right down to condemned prisoners with whom he often spent significant time and effort throughout his ministry. There are, within his *Journal*, also specific references to a range of people which can give some idea of the state of things and attitudes at the time:

Early in the *Journal* there is a detailed account of a tribe of the local Native Americans:

***Journal*, 20 Jul 1736:** Five of the Chicasaw Indians (twenty of whom had been in Savannah several days) came to see us, with Mr Andrews, their interpreter. They were all warriors, four of them head men. The two chiefs were Paustoobee and Mingo Mattaw. Our conference was as follows [*extracts only*]:

Q. Do you believe there is One above who is over all things?

Paustoobee answered, We believe there are four beloved things above: the clouds, the sun, the clear sky, and He that lives in the clear sky.

Q. Do you believe there is but One that lives in the clear sky?

A. We believe there are two with him, three in all.

Q. Do you think he made the sun, and other beloved things?

A. We cannot tell. Who hath seen?

Q. Do you think he made you?

A. We think he made all men at first.

Q. How did he make them at first?

A. Out of the ground.

Q. Do you believe he loves you?

A. I do not know. I cannot see him … From what I have seen.

When our enemies came against us before, then the beloved clouds came for us. And often much rain, and sometimes hail, has come upon them; and that in a very hot day. And I saw when many French, and Choctaws, and other nations came against one of our towns; and the ground made a noise under them, and the beloved ones in the air behind them; and they were afraid, and went away, and left their meat and drink, and their guns. I tell no lie. All these saw it too …

Q. Do you often think and talk of the beloved ones?

A. We think of them always, wherever we are. We talk of them and to them, at home and abroad; in peace, in war, before and after we fight; and, indeed, whenever and wherever we meet together.

Q. Where do you think your souls go after death?

A. We believe the souls of red men walk up and down, near the place where they died, or where their bodies lie; for we have often heard cries and noises near the place where any prisoners had been burned …

Q. How came your nation by the knowledge they have?

A. As soon as ever the ground was sound and fit to stand upon, it came to us, and has been ever since. But we are young men; our old men know more: but all of them do not know. There are but a few whom the beloved one chooses from a child, and is in them, and takes care of them, and teaches them: They know these things; and our old men practise; therefore they know. But I do not practise; therefore I know little.

***Journal*, 1 Jan 1737:** Our provisions fell short, our journey being longer than we expected; but having a little barbecued bear's flesh, (that is, dried in the sun,) we boiled it, and found it wholesome food. The next day we reached Darien, the settlement of the Scotch Highlanders: A sober, industrious, friendly, hospitable people.

On Monday evening we left Darien, and on Wednesday, came

to Frederica. Most here were, as we expected, cold and heartless: we found not one who retained his first love.

In the long description of the geography of the area Wesley learned about in Georgia there are fairly detailed descriptions of the local Native American tribes and where they lived: the Choctaws, Chicasaws, Cherikees and Uchees. The information has obviously been generally gleaned second-hand from reports of others and is generally sweepingly critical of many of the tribes: 'They are all, except, perhaps, the Choctaws, gluttons, drunkards, thieves, dissemblers, liars, implacable, unmerciful, murderers …'

Such attitudes by no means reflect Wesley's normal attitudes to local peoples.

***Journal*, 15 Oct 1759:** I walked up to Knowle, a mile from Bristol, to see the French prisoners. About eleven hundred of them we were informed, were confined in that little space without any thing to lie on but a little dirty straw, or any thing to cover them but a few foul thin rags, either by day or night, so that they died like rotten sheep. I was much affected, and preached in the evening on, [*Exodus 23:9*] 'Thou shalt not oppress a stranger; for ye know the heart of a stranger, seeing ye were strangers in the land of Egypt.'

Eighteen pounds were contributed immediately, which were made up four and twenty next day. With this we bought linen and woollen cloth, which were made up into shirts, waistcoats, and breeches. Some dozen of stockings were added; all of which were carefully distributed, where there was the greatest want. Presently after, the Corporation of Bristol sent a large quantity of mattresses and blankets. And it was not long before contributions were set on foot at London, and in various parts of the kingdom; so that I believe from this time they were pretty well provided with all the necessaries of life.

***Journal*, 16 Jul, 1760:** I took my leave of them at Clare in the afternoon, and in the evening returned to Limerick.

I rode to Newmarket, which was another German settlement. But the poor settlers, with all their diligence and frugality, could not procure even the coarsest food to eat, and the meanest raiment to put on, under their *merciful* landlords: So that most of these, as well as those at Ballygarane, have been forced to seek bread in other places; some of them in distant parts of Ireland, but the greater part in America.

***Journal*, 13 Oct 1790:** We got to Norwich between two and three. In the way we read Captain Carrel's *Travels*, admirably well wrote, and giving, I believe, a just account of the interior parts of North-America. Here is no gay account of the Islands of Pelew or Lapita, but a plain relation of matter of fact. Surely eastern and western savages are much alike; and some good might be found in the east as well as the west. But to see nature in perfection, either at Pelew, or elsewhere, we need only to look on the savages at Fort-William Henry, butchering, in cold blood, so many hundreds of helpless, unresisting men, in the very spirit of the old murderer.

Wesley generally preached at five in the morning. However, occasionally he also preached at ten 'for the sake of the gentry' but found even this hour could still be too early for them.

This final section gives extracts from Wesley's writings dealing with a human rights issue that was very dear to his heart and on which he had strong views – the widespread practice of slavery.

***Journal*, 10-12 Feb 1772:** In going to Dorking, I read Mr Jones's ingenious tract, upon Clean and Unclean Beasts. He really seems to prove his point, to make it reasonably plain, that there is a deeper design in that part of the Levitical Law, than is commonly understood: That God had a view throughout, to the moral, rather than natural, qualities of creatures which he pronounced unclean; and intended it as a standing warning to his people, against fierceness, greediness, and other ill properties, which so eminently belonged to those beasts or birds that they were forbidden to touch ...

In returning I read a very different book, published by an honest Quaker (Anthony Benezet), on that execrable sum of all villainies, commonly called the Slave trade. I read nothing like it in the heathen world, whether ancient or modern: And it infinitely exceeds, in every instance of barbarity, whatever Christian slaves suffer in Mahometan countries.

***Journal*, 13 Nov 1776:** I set out with Mr Fletcher to Norwich. I took coach at twelve, slept till six, and then spent the time very agreeably, in conversation, singing, and reading. I read Mr Bolt's account of the affairs in the East Indies: I suppose much the best that is extant. But what a scene is here opened! What consummate villains, what devils incarnate, were the managers there! What utter strangers to justice, mercy, and truth; to every sentiment of humanity! I believe no heathen history contains a parallel. I remember none in all the annals of antiquity; Not even the divine Cato, or the virtuous Brutus, plundered the provinces committed to their charge with such merciless cruelty as the English have plundered the desolated provinces of Indostan.

***Journal*, 14 Apr 1777:** I preached about noon at Warrington, and in the evening at Liverpool; where many large ships are now laid up in the docks, which has been employed for many years in buying or stealing poor Africans, and selling them in America for slaves. The men-butchers have now nothing to do at this laudable occupation. Since the American war broke out, there is no demand for human cattle. So the men of Africa, as well as Europe, may enjoy their native liberty.

However, the slave trade was by no means really over and one of the later of Wesley's published letters, sent from Guernsey, deserves to be reprinted in full. In his 10,000 word tract, *Thoughts upon Slavery*, which he refers to in this letter, he comprehensively surveys all aspects of the subject in quite appalling detail.

> **Letter, 18 Aug 1787:** GENTLEMEN, A week or two ago I was favoured with a letter from Mr Clarkson, informing him of his truly Christian design, to procure, if possible, an Act of Parliament for the abolition of slavery in our Plantations. I have long wished for the rolling away of this reproach from us, a reproach not only to religion, but to humanity itself. Especially when I read Mr Benezet's tracts, and what Mr Sharp has written upon the subject. My friends in America are of the same mind. They have already emancipated several hundred of the poor negroes, and are setting more and more at liberty every day, as fast as they can do it with any tolerable inconvenience. This is making a little stand against his shocking abomination; but Mr Clarkson's design strikes at the root of it. And if it can be put in execution will be a lasting honour to the British nation. It is with great satisfaction that I learn that so many of you are determined to support him. But without doubt, you [*may*] expect to meet with rough and violent opposition. For the slave holders are a numerous, a wealthy, and consequently a very powerful body. And when you bring their craft into danger, do you not touch the apple of their eye? Will they not then raise all their forces against you and summon their friends from every side? And will they not employ hireling writers in abundance, who will treat you without either justice or mercy? But I trust, Gentlemen, you will not be affrighted at this: no, not when some of your friends turn against you: perhaps some who have made the warmest professions of goodwill, and the strongest promises of assisting you. I trust you will not be discouraged thereby; but rather more resolute and determined. I allow, with men this is impossible; but we know all

things are possible with God! What little I can do to promote this excellent work I will do with pleasure. I will print a large edition of the tract I wrote some years since, *Thoughts upon Slavery*, and send it (which I have the opportunity of doing once a month) to all my friends in Great Britain and Ireland; adding a few words in favour of your design, which I believe will have some weight with them. I commend you to Him who is able to carry you through all opposition and support you in all discouragements, and am, Gentlemen

Your hearty well-wisher.

He also wrote several other letters to influential people strongly supporting the anti-slavery movement:

To end this chapter on a lighter note – Wesley's letters give encouragement to what was then currently considered the medicinal advantages of sea bathing:

Letter, Dewsbury, 10 Jul 1772: If I can I will contrive that you may be in a circuit which will give you convenience for bathing in the sea, though a river is as good.

To his niece, Sarah Wesley:

Letter, Bristol, 1 Sep 1788: My dear Sally, I received yours yesterday in the afternoon. As Ramsgate is more private, I am not sorry that you are there, and that you have so suitable a companion.

I think it would be expedient for you to bathe every day, unless you find yourself chilled when you come out. But I do not advise you to drink any sea water. I am persuaded it was never designed to enter any human body for any purpose but to drown it.

Letter, Bristol, 17 Sep 1790: My dear Sally, Will it not be best for you to spend a little time at Margate as soon as possible? … I believe sea-bathing will brace your nerves; but I pray [you do not drink the] sea-water.

6
Descriptive geographer

Towns and cities

During his travels John Wesley visited a great number of places and, being a geographer at heart, he often recorded descriptions in his *Journal*. He did not do this on a formal regular basis but just when he was particularly struck by certain aspects of where he was visiting. In many instances the contrast with modern times is most striking.

This section gives many of the more interesting impressions of towns and cities listed in alphabetical (not chronological) order by country. In a few cases interesting comments made on more than one visit are included.

England

Boston: About one I preached at Sibsley, on the edge of the Fens. There were a few wild colts here also; but all the rest (and they were not a few) were serious and deeply attentive. So were most of the congregation even at Boston, though much astonished, as not being used to field preaching.

I took a walk through the town. I think it is not much smaller than Leeds; but, in general, it is far better built. The church is indeed a fine building. It is larger, loftier, nay, rather more lightsome, than even St Peter's at Norwich; and the steeple is, I suppose, the highest tower in England, nor less remarkable for the architecture than the height. [Journal, *13 Aug 1761*]

*

We went on to Boston, the largest town in the county, except Lincoln. From the top of the steeple (which I suppose is by far the highest tower in England) we had a view not only of all the town, but of all the adjacent country. Formerly this town was in the fens; but the fens are vanished away: great part of them is turned into pasture, and a part into arable land. [Journal, *16 Jun 1780*]

Chester: We walked round the walls of the city, which are something more than a mile and three quarters in circumference. But there are many vacant spaces within the walls, many gardens, and a good deal of pasture ground: So that I believe Newcastle-upon-Tyne, within the walls, contains at least a third more houses than Chester.

The greatest convenience here is what they call 'the Rows'; that is, covered galleries, which run through the main streets on each side, from east to west, and from north to south; by which means one may walk both clean and dry in any weather, from one end of the city to the other. [Journal, *22 Jun 1752*]

Colchester: I rode on, through an extremely pleasant and fruitful country to Colchester. I have seen very few such towns in England. It lies on the ridge of an hill, with other hills on each side which lie parallel with it, at a small distance. The two main streets, one running east and west, and the other north and south, are quite straight, the whole length of the town, and full as broad as Cheapside.

I preached at four on St John's Green, at the side of an old high wall, (a place that seemed to be made on purpose,) to an extremely attentive audience. [Journal, *27 Oct 1758*]

Grimsby: I set out for the eastern part of Lincolnshire, and after preaching at Awkborough and Barrow in the way, came the next day to our old friends at Grimsby. It put me in mind of Purrysburg in Georgia. It was one of the largest towns in the county: It is no bigger than a middling village, containing a small number of half-starved inhabitants, without any trade, either foreign or domestic. But this they have; they love the Gospel, hardly six families excepted. [Journal, *18 Apr 1766*]

Hull: When I landed at the quay at Hull, it was covered with people enquiring, 'Which is he? Which is he?' But they only stared and laughed; and we walked unmolested to Mr A–'s house.

I was quite surprised at the miserable condition of the fortifications; far more ruinous and decayed than those at Newcastle, even before the rebellion. It was well there is no enemy near.

I went to Prayers at three in the old church, – a grand and venerable structure. [Journal, *24 Apr 1752*]

Lynn: [*Kings Lynn*] Lynn seems considerably larger than Yarmouth: I believe it stands on double the grounds; and the houses in general are better built: some of them are little palaces. The market-place is a spacious and noble square, more beautiful than either that at Yarmouth or Norwich; and the people are quite another turn, affable and humane. They have the openness and frankness common throughout the county; and they add to it good nature and courtesy. [Journal, *7 Nov 1771*]

Liverpool: At six in the morning, *Tuesday* 15, I preached to a large and serious congregation [*at Warrington*]; and then went on to Liverpool, one of the neatest, best-built towns I have seen in England: I think it is full twice as large as Chester; most of the streets are quite straight. Two-thirds of the town, we were informed, have been added within these forty years. If it continue to increase in the same proportion, in forty years more it will nearly equal Bristol. The people in general are the most mild and courteous I ever saw in a sea-port town, as indeed appears by their friendly behaviour, not only to the Jews and Papists who live among them, but even the Methodists (so called). [Journal, *14 Apr 1755*]

Oxford: Walking through the city, I observed it swiftly improving in everything but religion. Observing narrowly the Hall at Christ-Church, I was convinced it is both loftier and larger than that of the Stadt-House in Amsterdam. I observed also, the gardens and walks in Holland, although extremely pleasant, were not to be compared with St John's, or Trinity gardens; much less with the parks, Magdalen water-walks, &c., Christ-Church meadow, or the White-walk. [Journal, *15 Jul 1783*]

Portsmouth: We took horse again at half an hour past four, and reached Portsmouth about one.

After a little rest, we took a walk round the town, which is regularly fortified; and is, I suppose, the only regular fortification in Great Britain or Ireland. Gosport, Portsmouth, and the Common, (which is now all turned into streets,) may probably contain half as many people as Bristol; and so civil a people I never saw before in any seaport in England.

I preached at half an hour after six, in an open part of the Common adjoining the new church. The congregation was large and well behaved. [Journal, *9 Jul 1753*]

*

I walked round the Dock, much larger than any other in England. The late fire began in a place where no one comes, just at low water, and at the time all were fast asleep. So that none can doubt its being done by design. It spread with such amazing violence, among tow, and cordage, and dry wood, that none could come near without the utmost danger. Nor was anything expected but that the whole dock would be consumed, if not the town also. But this God would not permit. It stopped on one side, close to the Commissioner's house; and just as it was seizing the town on the other side, the wind changed and drove it back. Afterwards the fury of it was checked, by water, by sand, and by pulling down some buildings. And yet it was full five weeks before it was wholly put out. [Journal, *12 Oct 1770*]

Robinhood's Bay: I rode to Robinhood's Bay, near Whitby. The town is very remarkably situated: it stands close to the sea, and is in great part built on craggy and steep rocks, some of which rise perpendicular from the water. And yet the land, both on the north, south, and west is fruitful and well cultivated. I stood on a little rising near the quay, in a warm, still evening, and exhorted a multitude of people, from all parts, to 'seek the Lord, while he may be found.' They were all attention and most of them met me again at half an hour after four in the morning. [Journal, *8 May 1753*]

Sheerness: Such a town as many of these live in, is scarce to be found again in England. In the dock adjoining the fort there are six old men-of-war. These are divided into small tenements, forty, fifty, or sixty in a ship, with little chimneys and windows; and each of these contain a family. In one of them, where we called, a man and his wife and six children lived. And yet the ship was sweet and tolerably clean; sweeter than most ships I have been in. [Journal, *16 Jan 1768*]

South-Petherton: About noon I preached at Cathanger, about eight miles from Taunton … I preached in the great hall, like that of Lincoln College, to a very serious congregation. In the evening I preached at South-Petherton, once a place of renown, and the capital of a Saxon kingdom; as is vouched by a palace of King Ina still remaining, and a very large and ancient church. I suppose the last blow given to it was by Judge Jeffries, who, after the Monmouth rebellion, hanged so many of the inhabitants, and drove so many away, that it is never likely to lift up its head again. [Journal, *3 Sep 1778*]

Stourport: I went to Stourport. Twenty years ago there was but one house here; now there are two or three streets; and, as the trade swiftly increases, it will probably grow into a considerable town. A few years since Mr Cowell largely contributed to the building of a preaching-house here, in which both Calvinists and the Arminians might preach; But when it was finished, the Arminian preachers were totally excluded. Rather than go to law, Mr Cowell built another House, both larger and more convenient. I preached there at noon to a large congregation, but to a much larger in the evening. [Journal, *20 Mar 1788*]

Wincanton: I preached at Pensford and Shepton-Mallet in my way to Wincanton, one of the dullest places in all the county. I preached on Death in the evening, and Hell in the morning. It seemed, these were the very subjects they wanted. I never saw this careless people so much affected before. [Journal, *8 Oct 1770*]

Winchelsea: I walked over to Winchelsea, said to have been once a large city, with an abundance of trade and of inhabitants, the sea washing the foot of the hill on which it stands. The situation is exceeding bold, the hill being high and steep on all sides. But the town is shrunk almost into nothing, and the seven churches into half and one. I preached at eleven in the new Square, to a considerable number of serious people. [Journal, *30 Oct 1771*]

*

I went over to that poor skeleton of ancient Winchelsea. It is beautifully situated on top of a steep hill, and was regularly built in broad streets, crossing each other and encompassing a very large Square; in the midst of which was a large church now in ruins. I stood under a large tree, on the side of it, and called to most of the inhabitants of the town, 'The kingdom of heaven is at hand: repent, and believe the Gospel.' It seemed as if all that heard were, for the present, almost persuaded to be Christians. [Journal, *7 Oct 1790*]

The tree at Winchelsea under which Wesley preached his last sermon in the open air

Yarmouth: We went to Yarmouth, a cold, dead, uncomfortable place. I laboured to gather up the fragments of the poor society … In the evening I strongly exhorted them to 'repent and do the first works.' *Saturday* We returned to Norwich. [Journal, *2 Nov 1769*]

Scotland

Dumfries: Today we went on through lovely roads to Dumfries. Indeed all the roads are wonderfully mended since I last travelled this way. Dumfries is beautifully situated; but as to wood and water, and gently-rising hill, &c., is, I think, the neatest, as well as the most civilised, town that I have seen in the kingdom …

Near Dumfries there are five very large public gardens, which furnish the town with greens and fruit in abundance. [Journal, *13-14 May 1788*]

Dunkeld: From Methuen we went to Dunkeld, once the capital of the Caledonian kingdom; now a small town, standing on the bank of the Tay, and at the foot of several rough, high mountains ... We walked through the Duke of Athol's gardens, in which one thing I never saw before – a summer-house in the middle of a green-house, by means of which one might in the depth of winter enjoy the warmth of May, and sit surrounded with greens and flowers on every side. [Journal, *27-28 Apr 1772*]

Edinburgh: I took my leave of Edinburgh for the present. The situation of the city, on a hill shelving down on both sides, as well as to the east, with the stately castle upon a craggy rock on the west, is inexpressibly fine. And the main street, so broad and finely paved, with the lofty houses on either side, (many of them seven or eight stories high,) is far beyond any in Great Britain. But how can it be suffered that all manner of filth should still be thrown even into this street continually? Where are the Magistracy, the Gentry, the Nobility of the land? Have they no concern for the honour of their nation? How long shall the capital city of Scotland, yea, and the chief street of it, stink worse than a common-sewer? Will no lover of his country, or of decency and common sense, find a remedy for this?

Holyrood-House, at the entrance of Edinburgh, the ancient Palace of the Scottish Kings is a noble structure. It was rebuilt and furnished by King Charles the Second. One side of it is a picture gallery, wherein are pictures of all the Scottish Kings, and an original one of the celebrated Queen Mary. It is scarce possible for any who looks at this to think her such a monster as some have painted her; nor indeed for any who considers the circumstances of her death, equal to that of an ancient martyr.

I preached in the evening at Musselburgh, and at five in the morning. [Journal, *11 May 1761*]

Glasgow: We dined at a village called Lesmahaggy, and about eight in the evening reached Glasgow … I walked over the city, which I take to be as large as Newcastle-upon-Tyne. The University (like that of Dublin) is only one College, consisting of two small squares; I think not larger, nor at all handsomer, than those of Lincoln College, in Oxford. The habit of the students gave me surprise. They wear scarlet gowns, reaching only to their knees. Most I saw were very dirty, some very ragged, and all of a very coarse cloth. The high church is a fine building. The outside is equal to that of most cathedrals in England; but it is miserably defaced within; having no form, beauty, or symmetry left. [Journal, *17-18 Apr 1753*]

Greenock: I went on to Greenock, a sea-port town, twenty miles west of Glasgow. It is built very much like Plymouth-Dock, and has a safe and spacious harbour. The trade and inhabitants, and consequently the houses, are increasing swiftly; and so is cursing, swearing, drunkenness, sabbath-breaking, and all manner of wickedness. Our Room is about thrice as large as that at Glasgow; but it would not near contain the congregation. I spoke exceeding plain, and not without hope that we may see some fruit, even among this hard-hearted generation. [Journal, *20 Apr 1772*]

Inverness: After Edinburgh, Glasgow, and Aberdeen, I think Inverness is the largest town I have seen in Scotland. The main streets are broad and straight; the houses mostly old, but not very bad, nor very good. It stands in a pleasant and fruitful country, and has all things needful for life and godliness. The people in general speak remarkably good English, and are of a friendly, courteous behaviour. [Journal, *11 Jun 1764*]

St Andrew's: I paid a visit to St Andrew's, once the largest city in the kingdom. It was eight times as large as it is now, and a place of very great trade: But the sea rushing from the north-east, gradually destroyed the harbour and trade together: In consequence of which, whole streets (that were) are now meadows and gardens. Three broad, straight, handsome streets remain, all pointing at the old cathedral; which by the ruins appears to have been above three hundred feet long, and proportionably broad and high: So that it seems to have exceeded York Minster, and to have at least equalled any cathedral in England. Another church, afterwards used instead, bears date 1124. A steeple, standing near the cathedral, is thought to have stood thirteen hundred years.

What is left of St Leonard's College is only a heap of ruins. Two colleges remain. One of them has a tolerable square; but all the windows are broke, like those of a brothel. We were informed the students do this before they leave the College. Where are their blessed Governors in the meantime? Are they all fast asleep? The other college is a mean building, but has a handsome library newly erected. In the two Colleges, we learned, were about seventy students; near the same number as Old-Aberdeen. [Journal, *27 Jun 1776*]

Wales

Bangor to Conway: Having hired horses for Chester, we set out about seven. Before one we reached Bangor, the situation of which is delightful beyond expression. Here we saw a large and handsome cathedral, but no trace of the good old monks of Bangor; so many hundreds of whom fell a sacrifice at once to cruelty and revenge.

The country from thence to Penmaen Mawr is far pleasanter than any garden. Mountains of every shape and size, vales clothed with grass or corn, woods and smaller tufts of trees, were continually varying on the one hand, as was the sea prospect on the other. Penmaen Mawr itself rises almost perpendicular to an enormous height from the sea. The road runs along the side of it, so far above the beach, that one could not venture to look down, but that there is a wall built all along, about four foot high. Meantime the ragged cliff hangs over one's head, as if it would fall every moment.

An hour after we had left this awful place, we came to the ancient town of Conway. It is walled round; and the walls are in tolerably good repair. The castle is the noblest ruin I ever saw. It is four-square, and has four large round towers, one at each corner, the inside of which have been stately apartments. One side of the castle is a large church, the windows and arches of which are curiously wrought. An arm of the sea runs round two sides of the hill on which the castle stands; once the delight of Kings, now overgrown with thorns, and inhabited by doleful birds only. [Journal, *13 Aug 1756*]

Isle of Man

Castletown: A chaise was provided to carry me to Castletown. I was greatly surprised at the country. All the way from Douglas to Castletown it is as pleasant and as well cultivated as most parts of England, with many gentlemen's seats. Castletown a good deal resembles Galway; only it is not so large. At six I preached near the castle, I believe, to all the inhabitants of the town. [Journal, *31 May 1777*]

Douglas: After preaching in the evening, I went on board about eight o'clock, and before eight in the morning landed at Douglas, in the Isle of Man.

Douglas exceedingly resembles Newlyn in Cornwall; but in its situation, form, and buildings; only it is much larger, and has a few houses equal to most in Penzance. [Journal, *30-31 May 1777*]

Peel and general comments: A more loving, simple-hearted people than this I never saw. And no wonder; for they have but six Papists and no Dissenters, in the island. It is supposed to contain near thirty thousand people, remarkably courteous and humane. Ever since smuggling was suppressed, they diligently cultivate their land; And they have a large herring fishery, so that the country improves daily.

The old Castle at Peel (as well as the cathedral built within it) is only a heap of ruins. It was very large and exceeding strong, with many brass guns; but they are now removed to England. [Journal, *2 Jun 1777*]

Channel Islands

Wesley visited the Channel Islands and churches were set up there. He was obviously keen to spread the ministry as far as possible:

Dear Adam, You have reason to praise God for giving you such favour in the eyes of the poor people of Alderney. And I am in hopes our brother de Queteville [*the preacher from Guernsey*] will meet with a blessing in watering the seed which is already sown. But I observe in the map the name of another island, not far from Alderney. Are there none that understand English in the Isle of Sark? [*Letter, Birmingham, 26 Mar 1787*]

Ireland

Armagh: I preached at nine in the avenue at Armagh, to a large and serious congregation. It was increased four-fold at six in the evening; but many were there who behaved as if they had been in a bear-garden.

I took a walk to the Primate's; and went through the house and all the improvements. The house is neat and handsome, but not magnificent; and is elegantly, but not splendidly furnished.

The domain is beautifully laid out in meadow-ground, sprinkled with trees; on one side of which is a long hill covered with a shrubbery, cut into serpentine walks. On each side of the shrubbery is a straight walk, commanding a beautiful prospect. Since this Primate came, the town wears another face: he has repaired and beautified the cathedral, built a row of neat houses for the Choral-Vicars, erected a public Library and an Infirmary, procured the free-School to be rebuilt of the size of a little College, and a new-built horse-barrack, together with a considerable number of convenient and handsome houses; so that Armagh is at length rising out of its ruins into a large populous city. So much good may any man of a large fortune do, if he lays it out to best advantage! [Journal, *22 Jun 1778*]

Belfast: I rode in the afternoon to Belfast, the largest town in Ulster. Some think it contains near as many people as Limerick: It is far cleaner and pleasanter. [Journal, *23 Jul 1756*]

*

Thence we went to Belfast, the largest town in Ulster, said to contain thirty thousand souls. The streets are well laid out; are broad, straight, and well built. The poor-house stands on an eminence, fronting the main street, and having a beautiful prospect on every side over the whole country: The old men, the old women, the male and the female children, are all employed according to their strength; and all their apartments are airy, sweet, and clean, equal to any thing of the kind I have seen in England.

I preached in the evening on one side of the new church, to far the largest congregation I have seen in Ireland. [Journal, *9 Jun 1778*]

Clonmell: I went forward to Clonmell, the pleasantest town, beyond all comparison, which I have yet seen in Ireland. It has four broad, straight streets of well-built houses, which cross each other in the centre of the town. Close to the walls, on the south side, runs a broad, clear river. Beyond this rises a green and fruitful mountain, and hangs over the town. The vale runs many miles both east and west, and is well cultivated throughout. [Journal, *10 May 1756*]

Cork: In the evening I preached in the new House, at Cork, very near as large as that in Dublin; and far better finished in every respect, though at four hundred pounds less expense.

Walking up the Red-House Walk, (which runs between two rows of meadows, with the river winding through them, and a chain of fruitful hills on the right hand and on the left) I saw the plain reason why strangers usually complain of the unwholesomeness of the water in Cork. Many women were filling vessels with river-water (which is that commonly used in the city for tea and most other purposes) when the tide was at its height. Now although, this is not salt, yet it cannot but affect both the stomach and bowels of tender persons. [Journal, *12 and 17 May 1756*]

*

We went on to Cork. The latter was pleasant beyond description. At a very small distance on the left hand, the river 'rolled its sinuous train'; beyond which were shady trees, covering a steep hill and rising row above row. On the right we had another sloping mountain, tufted over with trees, sometimes forming one green, even wall, sometimes scattered up and down. Between these appeared several beautiful seats, some of them fit for Noblemen.

At six in the evening the preaching-house would ill contain all the congregation; and many rich and honourable were among them! Who had warned these to flee from the wrath to come? [Journal, *5 May 1787*]

Donegal: I took my leave of Londonderry … At five we reached Donegal, the county-town. What a wonderful set of county-towns are in this kingdom! Donegal and five more would not make up such a town as Islington. Some have twenty houses in them, Mayo three and Leitrim, I think, not one. Is not this owing in part to the fickleness of the nation, who seldom like anything long, and so are continually seeking new habitations, as well as new fashions, and new trifles of every kind.

We breakfasted at Ballyshannon, I believe the largest and pleasantest town in the county. [Journal, *27-28 May 1765*]

Downpatrick: After service we went to Downpatrick, where I slept in peace.

I walked through the town; I suppose one of the most ancient in Ulster. I was informed, it was once abundantly larger than it is now; consisting of the Irish town, then inhabited by none but Roman Catholics; and the English town, encompassed with a wall, and a deep ditch filled with water. At the head of the English town stands the Abbey on a hill which commands the country. It is a noble ruin, and is far the largest building that I have seen in the kingdom. Adjoining to it is one of the most beautiful groves which I ever beheld with my eyes: It covers the sloping side of the hill, and has vistas cut through it in every way. In the middle is a circular space, twenty of thirty yards in diameter. I would have preached there, but the rain drove into the House as many as could crowd together. [Journal, *11-12 Jun 1778*)

Galway: We took horse about ten, and rode through the fruitful and pleasant county of Galway. After having heard so much of the barrenness of this county, I was surprised, in riding almost the whole length of it, from south-east to north-west, to find only four or five miles of rocky ground, like the west of Cornwall; all the rest exceeded most that I have seen in Ireland. We came to Galway pretty well tired ... The town is old, and not ill-built, most of the houses being of stone, and several stories high. It is encompassed with an old, bad wall, and is in no posture of defence, either toward the land or toward the sea. Such is the supine negligence of both the English and Irish! [Journal, *25-26 Jun 1756*]

Kilkenny: We rode to Kilkenny, one of the pleasantest and the most ancient cities in the kingdom; and not inferior to any at all in wickedness, or in hatred to *this way*. I was therefore glad of a permission to preach in the Town-Hall; where a small, serious company attended in the evening.

I went to the Cathedral; one of the best-built which I have seen in Ireland. The pillars are all of black marble; but the late Bishop ordered them to be white-washed! Indeed, marble is so plentiful near this town, that the very streets are paved with it. At six in the evening I began preaching at the old Bowling-green near the Castle. [Journal, *11 Jul 1762*]

Newport: In the afternoon I rode over to Newport, eleven miles from Castlebar. About thirty years ago, a little company of Protestants settled here, by a river side, on the very extremity of the land, and built a small town. It has a fruitful hill on each side, and a large bay to the west, full of small fertile islands, containing one to several thousand acres. Of these they compute above three hundred, and near an hundred are inhabited; but by Papists alone, there not being a single Protestant among them! I went directly to the Rector's, who had before given me an invitation. Between seven and eight I preached to (I suppose) more than all the Protestants in the town. [Journal, *29 Jun 1756*]

Newry: No sooner did we enter Ulster than we observed the difference. The ground was cultivated just as in England; and the cottages not only neat, but with doors, chimneys, and windows. Newry, the first town we came to, (allowing for the size,) is built much after the manner of Liverpool. I preached soon after seven to a large congregation, and to a great part of them at five in the morning. Afterwards I spoke to the members of the society, consisting of Churchmen, Dissenters, and Papists that were. But there is no striving among them, unless to 'enter in at the strait gate.' [Journal, *20 Jul 1756*]

Portaferry: We had a pleasant ride to Portaferry, a pretty large sea-port town, and one of the quietest I ever saw, either in England, Scotland or Ireland. Here likewise I preached in a large meeting-house, to a serious and well-behaved congregation. [Journal, *9 Jun 1789*)

Prosperous: I went through a delightful country to Prosperous; a little town, begun five years ago by Captain Brooke, just returned from the East Indies. Here he introduced every branch of the cotton manufactory, on a most extensive plan. He built two rows of commodious houses, with all convenient appurtenances; and he now employs about a thousand men, women, and children, on the spot, beside near the same number in other places.

They had a very large Room, but not near large enough for the congregation. All that got in seemed much affected, as they did likewise at five in the morning. [Journal, *18 Apr 1785*]

Sligo: I rode through James's Town, once a strong place, now a heap of ruins, and through Carrick and Boyle, both enclosed by a pleasant and fruitful country. Soon after we entered the county of Sligo, the best peopled that I have seen in the kingdom. Eight villages we counted within seven miles: The town itself, I think, is little less than Limerick. The country is fertile and well-improved; even the mountains to the very top. It lies two miles from the sea, having a large harbour, covered by mountains on each side.

I preached in the evening, near the main street, to a small quiet, serious company. [Journal, *27 May 1758*]

Tanderagee: At eleven I preached in the church-yard at Lurgan ... In the afternoon I went on to Tanderagee, one of the pleasantest towns in Ireland, surrounded by woods and fruitful hills, with a clear river running between them. At six I stood in the Grove, where the tall elms shaded both me and the numerous congregation. Several gentlemen and several Clergymen were among them, and all behaved with serious attention. [Journal, *12 Jun 1785*]

Wesley's farewell to Ireland

Holland

Wesley travelled to Continental Europe both early and late in his ministry and recorded his varied impressions in his *Journal* and letters which included more specific and very diverse details of some of the places in Holland he visited:

Amsterdam: At six in the morning we took boat. The beautiful gardens lie on both sides the river, for a great part of the way to Amsterdam, whither we came about five in the evening. The exact neatness of all the buildings here, the nice cleanness of the streets, (which we were informed, were all washed twice a week,) and the canals which run through all the main streets, with rows of trees on either side, make this the pleasantest city which I have ever seen. [Journal, *17 Jun 1738*]

Colen: Yesterday evening we came hither. It is the ugliest, dirtiest town I ever yet saw. There is neither form nor comeliness belonging to it. The great church itself is mere heaps upon heaps – a vast misshaped or rather no-shaped building, with no regularity or proportion within or without; many of the stones broken, the windows dusty and full of cobwebs, and the pavement less clean than that of many English stables. [*Letter to his mother Jun 28 1738*]

Delft: At eleven we came to Delft, a large, handsome town, where we spent an hour at a merchant's house, who, as well as his wife, a very agreeable woman, seemed both to fear and to love God. Afterwards we saw the great church, I think nearly, if not quite, as long as York Minster. It is exceedingly light and elegant within, and every part is kept exquisitely clean. The tomb of William the First is much admired; particularly his statue, which has more life than one would think could be expressed in brass. [Journal, *16 Jun 1783*]

Haerlem: We took boat for Haerlem. The great church here is a noble structure, equalled by few cathedrals in England, either in length, breadth, or height: the organ is the largest I ever saw, and is said to be the finest in Europe … Before dinner we took a walk in Haerlem Wood. It adjoins the town, and is cut out in many shady walks; with lovely vistas shooting out in every way. The walk from The Hague to Scheveling is pleasant; those near Amsterdam more so; but these exceed them all … We returned in the afternoon to Amsterdam; and in the evening took leave of as many of our friends as we could. How entirely we were mistaken about the Hollanders, supposing them to be of a cold, phlegmatic, unfriendly temper! I have not met with a more warmly affectionate people in all Europe! [Journal, *25-26 Jun 1783*]

The Hague: When we came to The Hague, though we had heard much of it, we were not disappointed. It is, indeed, beautiful beyond expression. Many of the houses are exceeding grand, and are finely intermixed with water and wood; yet not too close, but so as to be sufficiently ventilated by the air …

As we walked over the Palace we saw the Swiss Guards at their exercise. They are a fine body of men, taller, I suppose, than any English regiment; and they all wear large whiskers, which they take care to keep as black as their boots. Afterwards we saw the gardens at the Old Palace, beautifully laid out, with a large piece of water in the middle, and a canal at each end: The open walks in it are pleasant, but the shady serpentine walks are far pleasanter. [Journal, *16-17 Jun 1783*]

Rotterdam: Never did common fame more grossly vary from the truth than in the English accounts of Holland. They tell us of a dirty, slovenly, unpolished people, without good nature, good manners, or common decency; whereas the very first thing that must strike every one that has eyes, and that before he has gone a hundred yards from Rotterdam Haven, is that this is the cleanest place he ever saw in his life, there bring scarce a speck of dirt to be seen either on the doors or steps of any of the houses or on the stones of the street. And all of the natives he meets, whether men, women, or children, are of a piece with the place they live in … Every one is ready to show you the way, or to answer any questions, without anything of English surliness. [*Letter to his mother, 19 Jun 1738*]

*

At Briel we took a boat for Rotterdam …We took a walk together round the town, all as clean as a gentleman's parlour. Many of the houses are as high as those in the main street at Edinburgh; and the canals, running through the chief streets, make them convenient as well as pleasant; bringing the merchants' goods up to their doors. Stately trees grow on all their banks. The whole town is encompassed with a double row of elms; so that one may walk all round it in the shade … In the evening we again took a walk round the town, and I observed,

1. Many of the houses are higher than most in Edinburgh. It is true they have not so many stories; but each story is far loftier.

2. The streets, the outside and inside of their houses in every part, doors, windows, well-staircases, furniture, even floors, are kept so nicely clean that you cannot find a speck of dirt.

3. There is such a grandeur and elegance in the fronts of the large houses, as I never saw elsewhere; and such a profusion of marble within, particularly in their lower floors and staircases, as I wonder other nations do not imitate.

4. The women and children (which I least of all expected) were in general the most beautiful I ever saw. They were surprisingly fair, and had an inexpressible air of innocence in their countenance.

5. This was wonderfully set off by their dress, which was *simplex munditiis*, plain and neat in the highest degree.

6. It has lately been observed that growing vegetables greatly

resist putridity; so there is an use in their numerous rows of trees which was not thought of at first. The elms balance the canals, preventing the putrefaction which those otherwise might produce.

One little circumstance I observed, which I suppose is peculiar to Holland: To chamber-windows a looking glass is placed on the outside of the sash, so as to show the whole street, with all the passengers. There is something very pleasing in these moving pictures. Are they found in no other country? [Journal, *11-14 Jun 1783*]

Utrecht: Utrecht has much the look of an English town. The streets are broad, and have many noble house. In quietness and stillness it much resembles Oxford. The country all round is like a garden; And the people I conversed with are not only civil and hospitable, but friendly and affectionate, even as those at Amsterdam. [Journal, *29 Jun 1783*]

*

General Comments: We always looked upon the Dutch as an heavy, dull, stoical people. But truly most, nay I may say all, with whom I conversed familiarly were as tender-hearted and as earnestly affectionate as the Irish themselves. [Letter, *20 Jul 1783]*

If you go abroad, I would by no means advise you go to France. That is no place to save expense; but it is the only place to make your sons coxcombs and your daughters coquettes. I cannot but think there is no country in Europe which would answer your design so well as Holland; and no place in Holland so well as Utrecht. It is within a day's journey of Helvoetsluys, whence you go directly by the packet for England. It is an healthful and pleasant city, and less expensive than almost any city in France. There are schools for your children; and if you should choose it, an university for your sons. [Letter, *30 Aug 1784*]

Germany

Dresden: At two in the afternoon we came to Dresden, the chief city of Saxony. Here also we were carried for above two hours from one Magistrate or Officer to another, with the usual impertinent solemnity, before we were suffered to go to our inn ... In the evening we saw the palace the late Elector was building when God called him away. The stone-work he had very near finished, and some of the apartments within. It is a beautiful design; but all is now swiftly running to ruin. The new church on the outside resembles a theatre. It is eight-square, built of fine freestone. We were desired to take notice of the great bridge which joins the new with the old town; of the large, brass crucifix upon it, generally admired for the workmanship; and of the late King Augustus's statue on horseback, which is at a small distance from it. Alas! where will all these things appear, when the earth and the works thereof shall be burned up? [Journal, *30 Jul 1738*]

Hernhuth: At three in the afternoon I came to Hernhuth, about thirty English miles from Dresden. It lies in Upper Lusatia, on the border of Bohemia, and contains about a hundred houses, built on rising ground, with evergreen woods on two sides, gardens and cornfields on the others, and high hills at some distance. It has one long street, through which the great road from Zittau to Lobau goes. Fronting the middle of this street is the Orphan-house; in the lower part of which is the apothecary's shop; in the upper, the chapel, capable of containing six or seven hundred people. Another row of houses runs at a small distance from either end of the Orphan-house, which accordingly divides the rest of the town (besides the long street) into two Squares. At the east of this is the Count's house; a small plain building like the rest: having a large garden behind it, well laid out, not for show, but for the use of the community ... The burying-ground (called by them Gottes Acker, that is God's ground) lies a few hundred yards out of the town, under the side of a little wood. There are distinct squares in it for married men and unmarried, for married and unmarried women, for male and female children, and for widows. [Journal, *1 and 8 Aug 1738*]

Meissen: About seven in the morning we came to Meissen. In Meissen Castle, the German chinaware is made, which is full as dear as that imported from the Indies; and as finely shaped, and beautifully coloured, as any I have ever seen. I was greatly surprised at all I saw there: At the costliness of the apparel in many, and the gaudiness of it in more. [Journal, *30 Jul 1738*]

Georgia and its settlements

The geographical observer in Wesley came to the fore in December 1737 in the lengthy piece he wrote on the area of Georgia he had learned about. This piece of his *Journal* is so informative as to justify the inclusion of significant portions:

Georgia lies in the 30th and 31st degree of north latitude. The air is generally clear, the rains being much shorter, as well as heavier, than in England. The dews are very great ... The land is of four sorts – pine-barren, oak-land, swamp and marsh ...

Savannah stands on a flat *bluff* , (so they term any high-land hanging over a creek or river,) which rises forty-five feet perpendicular from the river, and commands it several miles both upwards and downward. The soil is a white sand for above a mile in breadth, south-east and north-west. Beyond this, eastward, is a river-swamp; westward a small wood, in which was the old Indian town. On the other side of the river is a marshy island, covered with large trees. South-west of the town is a large pine-barren, which extends backward to a branch of the Alatamahaw river.

St Simon's Island, having on the south-east the Gulf of Florida, on the other side, branches of the Alatamahaw, is about one hundred miles south of Savannah, and extends in length about twenty, in breadth from two to five miles. On the west side of it, on a low bluff, stands Frederica, having woods to the north and south; to the east, partly woods, partly savannah, and partly marshes. The soil is mostly blackish sand. There is not much pine-land on the island, the greatest part being oak-land, intermixed with many savannahs, and the old Spanish or Indian fields. On the sea-port, about five miles south-east of the town, is the fort where the soldiers are stationed. But the storehouse in Frederica better deserves that name; being encompassed with regular ramparts of earth, and a palisaded ditch, and mounted with cannon, which entirely command the river.

About twenty miles north-west of St Simon's is *Darien*, the settlement of the Scotch Highlanders, a mile from Fort King George, which was built about seventeen and abandoned about eleven years since. The town lies on the mainland, close to a branch of the Alatamahaw, on a bluff about thirty feet above the river, having woods on all sides. The soil is a blackish sand. They built at first many scattered huts; but last spring [*1736*] expecting the Spaniards, they built themselves a large fort, and retired within the walls of it.

Augusta, distant from Savannah one hundred and fifty miles, and five from old Savannah Town, is designed to stand in an old Indian field, on a bluff about thirty feet high. A small fort of wooden piles was built there in 1737.

Old-Ebenezer, where the Saltzburghers settled at first, lies twenty-five miles west of Savannah. A small creek runs by the town, down to the river, and many brooks run between the little hills: but the soil is hungry, barren sand; and upon any sudden shower, the brooks rise several feet perpendicular and overflow whatever is near them.

New-Ebenezer, to which the Saltzburghers removed in March 1736, lies six miles eastward from the old, on a high bluff, near the Savannah river. Here are some tracts of fruitful land, though the greatest part of that adjoining the town is pine-barren. The huts, sixty in number, are neatly and regularly built; the little piece of ground allotted to each for a garden is everywhere put to best use, no spot being left unplanted.

About ten miles east of this, on a creek, three miles from the river, was the village of Abercorn. Ten families settled here in 1733; but it is now without inhabitant.

Five miles south-west of Savannah, on a small rise, stands the village of *Highgate*. It has pine-land on three sides and swamp on the fourth. Twelve families were placed here in 1733; nine thereof remain there. A mile eastward of this is *Hampstead*, settled with twelve families also, a little before Highgate; five of which are still remaining.

Six miles south-east of Savannah is *Thunderbolt*: Three families are settled here, near a small, ruinous fort. Four miles south of this is the island of *Skidoway*: On the north-east point wherof ten families were placed in 1734; (a small fort was built here likewise;) but nine of them are either dead, or removed to other places. A small creek divides *Skidoway* from *Tybee-Island*, on the south-east part of which, fronting the inlet, the lighthouse is built. Ten families were settled here in 1734; but they are part dead, and part removed, so that the island is now again without any fixed inhabitant.

The southermost settlement in Georgia is *Fort St Andrew*. It stands fifty miles south of Frederica, on the south-west side of Cumberland Island, upon a high neck of land, which commands the river both ways. The walls are of wood filled up with earth, round which are a ditch and palisade.

Landscapes

In addition to his descriptions of cities, towns and villages Wesley also included in his *Journal* accounts of landscapes that he passed through that had caught his particular attention, normally for their attractiveness. These can be very evocative of the areas concerned and some have already been included within other passages elsewhere. Some further examples, spread through a range of countries visited, are included below to give further insight into his travelling life and experiences. It is, however, clear from his comments that he uses the term 'mountain' more frequently than would be justified by the modern understanding of the term. He also appears to be rather overgenerous with his assessments in giving top accolades to so many locations!

England

***Journal*, 29 Aug 1743:** About sunset we were in the middle of the first great pathless moor beyond Launceston. About eight we were got quite out of the way; but we had not gone far before we heard the Bodmin bell. Directed by this, we turned to the left, and came to the town before nine.

***Journal*, 6 May 1747:** I rode to Shore, four miles south from Stonesey, lying about half way down an huge, steep mountain … We then climbed up to Todmorden-Edge, the brow of a long chain of mountains … We left the mountains and came down to the fruitful valley of Rosendale.

***Journal*, 25 Apr 1755:** About ten I preached near Todmorden … One can hardly conceive any thing more delightful than the vale through which we rode from thence. The river ran through the green meadows on the right. The fruitful hills and woods rose on either hand: Yet here and there a rock hung over, the little holes of which put me in mind of those beautiful lines. [*of Psalm 104*]: 'The hills are a refuge for the wild goats, and so are the stony rocks for the conies.' At three in the afternoon I preached at Heptonstall, on the brow of a mountain.

***Journal*, 8 Sep 1769:** [*Near Bridgewater*] This afternoon I went to the top of Brent-Hill: I know not I ever before saw such a prospect. Westward one may see to the mouth of the Bristol Channel; and three other ways as far as the eyes can reach. And most of the land which you see is well cultivated, well wooded, and well watered: So that the globe of earth, in its present condition, can hardly afford a more pleasing scene.

***Journal*, 5 Dec 1770:** We went to Dover, where, with some difficulty, we climbed to the top of Shakspeare's Cliff. It is exceeding high, and commands a vast prospect both by sea and land; but it is nothing so terrible in itself as it is in his description.

***Journal*, 2 Jun 1772:** We rode to New Orygan, in Teesdale … From the top of the next enormous mountain, we had a view of Weardale. It is a lovely prospect. The green gently-rising meadows and fields, on both sides of the little river, clear as crystal, were sprinkled over with innumerable little houses; three in four of which (if not nine in ten) are sprung up since the Methodists came hither.

***Journal*, 12 Oct 1774:** The Isle (or properly Peninsula) of Purbeck is nine or ten miles broad, and perhaps twenty long, running nearly from north-east to south-west. Two mountains run nearly the whole length, with valleys between them and on each side, but poorly cultivated. The people in general are plain, artless, good-natured and well behaved.

***Journal*, 9 Jul 1777:** I went through a delightful vale to

Malvern-Wells, lying on the side of a high mountain, and commanding one of the finest prospects in the world; the whole vale of Evesham. Hitherto the roads were remarkably good; but they grew worse and worse until we came to Monmouth.

***Journal*, 3 Jun 1790:** We rode to Hexham, through one of the pleasantest countries that I have lately seen. The road lay (from Haisle) on the side of a fruitful mountain shaded with trees, and sloping down to a clear river; which ran between ours and another fruitful mountain, well wooded and improved.

Wales

***Journal*, 3-5 May 1743:** I set out for Wales, in company with one who was my pupil at Oxford. We could get that night no farther than the Bull, five Welsh miles beyond Abergavenny. The next morning we came to Builth ... I rode over such rugged mountains as I ever saw before, to Cardiff.

***Journal*, 5 Aug 1747:** Taking horse early in the morning, we rode over the rough mountains of Radnorshire and Montgomeryshire into Merionethshire. In the evening I was surprised with one of the finest prospects, in its kind, that ever I saw in my life. We rode in a green vale, shaded with rows of trees, which made an arbour for several miles. The river laboured along on our left hand, through broken rocks of every size, shape, and colour. On the other side of the river, the mountains rose to an immense height, almost perpendicular: And yet the tall straight oaks stood, rank above rank, from bottom to the very top; only here and there, where the mountain was not so steep, were interposed pastures or fields of corn. At a distance, as far as the eye could reach, as it were by way of contrast 'a mountain huge uprear'd its broad bare back' with vast, rugged rocks hanging over its brow, that seemed to nod portending ruin.

***Journal*, 20 Aug 1763:** We took horse at four, and rode through one of the pleasantest countries in the world. When we came to Trecastle, we had rode fifty miles in Monmouthshire and Brecknockshire; and I will be bold to say, all England does not afford such a line of fifty miles' length, for fields, meadows, woods, brooks, and gently rising mountains, fruitful to the very top. Carmarthenshire, into which we came soon after, has at least as fruitful soil; but it is not so pleasant, because it has fewer mountains, though abundance of brooks and rivers.

Scotland

***Journal*, 31 May 1757:** I breakfasted at Dumfries ... We then rode through an uncommon pleasant country (so widely distant is common report from truth) to Thorny-Hill, two or three miles from the Duke of Queensborough's seat; and ancient and noble pile of building, delightfully situated on the side of a pleasant and fruitful hill ... We rode afterward partly over and partly between some of the finest mountains, I believe in Europe; higher than most, if not any, in England, and clothed with grass to the very top. Soon after we came to Lead-Hill, a little town at the foot of the mountains, wholly inhabited by miners. We rode on to Glasgow.

Ireland

***Journal*, 9 Apr 1748:** I preached in Connaught, a few miles from Athlone ... The Shannon comes within a mile from the house where I preached. I think there is not such another river in Europe: It is here ten or twelve miles over, though scarce thirty miles from its fountain-head. There are many islands in it, once well inhabited, but now mostly desolate. In almost every one is the ruins of a church: In one, the remains of no less than seven.

***Journal*, 15 Jun 1750:** We set out, and reached Kilkenny, about twenty-five old Irish miles, about noon. This is by far the most pleasant, as well as most fruitful country, which I have seen in all Ireland. Our way after dinner lay by Dunmore, the seat of the late Duke of Ormond. We rode through the Park for about two miles, by the side of the river runs. I never saw either in England, Holland, or Germany, so delightful a place. The walks, each consisting of four rows of ashes, the tufts of trees sprinkled up and down, interspersed with the smoothest and greenest lawns, are beautiful beyond description.

***Journal*, 25 Jun 1756:** We took horse about ten, and rode through the fruitful and pleasant county of Galway. After having heard so much of the barrenness of this county, I was surprised, in riding almost the whole length of it, from south-east to north-west, to find only four or five miles of rocky ground, like the west of Cornwall; all the rest exceeded most that I have seen in Ireland.

***Journal*, 29 Aug 1765:** I was desired by some friends to take a ride to the Dargel, ten or twelve miles from Dublin; one of the greatest natural curiosities, they said, which the kingdom afforded. It far exceeded my expectation. You have a high and steep mountain, covered with stately wood, up the side of which a path is cut, and seats placed at small distances. A deep vale, through which a clear river runs, lies between this and another high mountain, whose sides are clothed with tall trees, row above row, from the river to the very top. Near the summit of the first mountain, you have an opening on the one hand which commands the fruitful counties of Kildare, Dublin, and Louth, as far, in a clear, sunshiny day, as the high mountains of Newry; on the other hand is a fine landscape of meadows and fields, that terminate in a sea-prospect. Adding this to the rest, it exceeds anything which I have seen in Great Britain.

Journal, 27 Jun 1789: We returned to Dublin by the Glen of the Downs, much resembling that which lies north above Keswick-Water. All this country is remarkably fruitful and pleasant, having, in many parts, a fine sea, as well as land, prospect.

Isle of Man

Journal, 5 and 6 Jun 1781: In the afternoon we rode through a pleasant and fruitful country, to Ramsay, about as large as Peel, and more regularly built ... This morning we rode through the most woody, and far pleasantest part of the island; a range of fruitful land, lying at the foot of the mountains, from Ramsay, through Sulby, to Kirkmichael.

Holland

Journal, 23 Jun 1783: After dinner Mrs J— took me in a coach to the Mere, and thence round the country to Zeeberg. I never saw such a country before: I suppose there is no such summer country in Europe. From Amsterdam to Mere is all a train of the most delightful gardens. Turning upon the left, you then open upon the Texel, which spreads into a sea. Zeeberg itself is a little house built on the edge of it, which commands both a land and sea prospect.

Parks and gardens

One of Wesley's most valued methods of relaxation was to visit and wander in parks and gardens whenever the possibility arose. This applied, not only when he was on his travels, but also when he was working at home as the following extract relates:

Journal, 20 Aug 1782: My brother and I paid our last visit to Lewisham, and spent a few pensive hours with the relict of our good friend, Mr Blackwell. We took one more walk round the garden and meadow, which he took so much pains to improve. Upwards of forty years this has been my place of retirement, when I could spare two or three days from London. In that time, first Mrs Sparrow went to rest; then Mrs Dewall; then good Mrs Blackwell; now Mr Blackwell himself. Who can tell how soon we may follow them?

As he grew more mature in years the appeal of parkland obviously increased, with his *Journal* containing more entries of this type. A varied selection of these very descriptive and mostly positive extracts follows which provide a pleasant ending for this book:

England

***Journal*, 13 Sep 1764:** [*Near Bristol*] I spent an hour in Lord B—'s gardens, or more properly, woods. They are small to the late Duke of Kent's in Bedfordshire, and, therefore, not capable of so much variety; but for the size, it is not possible for anything of the kind to be more agreeable; and the situation, on the top of an high hill, in one of the fruitfullest counties in England, gives them an advantage which even Stow-Gardens have not.

***Journal*, 11 Oct 1775:** [*Near Wycombe*] I took a walk to Lord Shelburne's house. What a variety, in so small a compass! A beautiful grove, divided by a serpentine walk, conceals the house from the town. At the side of this runs a transparent river, with a smooth walk on each bank. Beyond this is a level lawn; then the house with sloping gardens behind it. Above these is a lofty hill; near the top of which is a lovely wood, having a grassy walk running along, just within the skirts of it.

***Journal*, 12 Sep 1776:** [*Somerset*] I spent two hours in Mr Hoare's gardens, at Stourton. I have seen the most celebrated gardens in England; but these far exceed them all: 1. In the situation; being laid out on the sloping sides of a semicircular mountain: 2. In the vast basin of water inclosed between them, covering, I suppose, sixty acres of ground: 3. In the delightful interchange of shady groves and sunny glades, curiously mixed together. Above all in the lovely grottoes, two of which excel everything of the kind which I ever saw; the fountain-grotto, made entirely of rock-work, admirably well imitating nature; and the castle-grotto, into which you enter unawares beneath a heap of ruins. This is within totally built of roots of trees wonderfully interwoven. On one side of it is a little hermitage, with a lamp, a chair, a table, and bones upon it.

However, even these gardens, and also the gardens in the subsequent extract, did not fully meet with Wesley's tastes as he continues:

Others were delighted with the temples, but I was not: 1. Because several of the statues about them were mean: 2. Because I cannot admire the images of devils; and we know the gods of the Heathens are but devils: 3. Because I defy all mankind to reconcile statues with nudities, either to common sense or common decency.

In 1779 Wesley took the opportunity of his visit to Stow gardens to include in his *Journal* a detailed comparative critique of this and gardens he had already visited at Stourhead and Cobham:

***Journal*, 13 Oct 1779:** I now desired to take a view of the much more celebrated gardens at Stow. The first thing I observed was the beautiful water which runs through the gardens, to the front of the house. The tufts of trees, placed on each side of this, are wonderfully pleasant; and so are many of the walks and glades through the woods, which are disposed with a fine variety. The large pieces of water interspersed give a fresh beauty to the whole. Yet there are several things which must give disgust to any person of common sense: 1. The buildings, called Temples, are most miserable, many of them both within and without. Sir John Vanbrugh's is an ugly, clumsy lump, hardly fit for a gentleman's stable. 2. The temples of Venus and Bacchus, though large, have nothing elegant in the structure; and the paintings in the former, representing a lewd story, are neither well designed nor executed. Those in the latter are quite faded, and most of the inscriptions vanished away. 3. The statues are full as coarse as the paintings, particularly those of Apollo and the Muses, whom a person, not otherwise informed, might take to be nine cook-maids. 4. Most of the water in the ponds is dirty, and thick as puddle. 5: It is childish affectation to call things here by Greek or Latin names, as Styx, and the Elysian Fields. 6. It was ominous for My Lord to entertain himself and his noble company in the grotto built on the banks of Styx; that is, on the brink of hell. 7. The river on which it stands is a black, filthy puddle, exactly resembling a common sewer. 8. One of the stateliest monuments is taken down, – the Egyptian Pyramid; and no wonder, considering the two inscriptions, which are still legible [*translated from the Latin of the original*]:

'Thy lands, thy dome, thy pleasing wife,
These must thou quit; 'Tis nature's doom:
No tree whose culture charms thy life,
Save the sad cypress, waits thy tomb.

Of meat and drink, of pleasure and of play
You've had your fill: 'Tis time you steal away,
Lest you be jostled off by that brisk race
Whom sports become, and wanton revels grace.'

Upon the whole, I cannot but prefer Cobham gardens to those at Stow: For 1. The river at Cobham shames all the ponds at

Stow. 2. There is nothing at Stow comparable to the walk near the wheel which runs up the side of a steep hill, quite grotesque and wild. 3. Nothing in Stow gardens is to be compared to the large temple, the pavilion, the antique temple, the grotto, or the building at the head of the garden; nor to the neatness which runs through the whole.

But there is nothing even at Cobham to be compared, 1. To the beautiful cross at the entrance of Stourhead gardens. 2. To the vast body of water. 3. The rock-work grotto. 4. The temple of the sun. 5. The hermitage. Here too every thing is nicely clean, as well as in full preservation. Add to this, that all the gardens hang on the sides of a semicircular mountain. And there is nothing either at Cobham or Stow which can balance the advantage of such a situation.

***Journal*, 12-13 Jul 1782:** [*Near Birmingham*] I walked through Mr Bolton's curious works. He has carried everything which he takes in hand to a high degree of perfection, and employs in the house about five hundred men, women, and children. His gardens, running along the side of a hill, are delightful indeed; having a large piece of water at the bottom, in which are two well-wooded islands. If faith and love dwell here, then there may be happiness too. Otherwise all these beautiful things are as unsatisfactory as straws and feathers.

I spent an hour in Hagley-Park; I suppose inferior to few, if any, in England. But we were straitened for time. To take a proper view of it, would require five or six hours. Afterwards I went to the Leasowes, a farm so called, four or five miles from Hagley. I never was so surprised. I have seen nothing in all England to be compared with it. It is beautiful and elegant all over. There is nothing grand, nothing costly; no temples, so called; no statues; (except two or three, which had better have been spared;) but such walks, such shades, such hills, such dales, such lawns, such artless cascades, such waving woods, with water intermixed, as exceed all imagination! On the upper side, from the openings of a shady walk, is a most beautiful and extensive prospect. And all this is comprised in the compass of three miles! I doubt if it be exceeded by anything in Europe.

***Journal*, 16 Oct 1782:** I saw such a garden at Oxford as I verily believe all England cannot parallel. It is three-square; and, I conjecture, contains about an acre of ground: It is filled with fruit-trees of various sorts, and all excellent in their kinds. But it is odd beyond all description; superlatively whimsical. The owner has crowded together pictures, statues, urns, antiques of various kinds.

***Journal*, 24 Aug 1783:** I went with a few friends to Blase-Castle [*near Bath*]. The woods on the side of the hill, cut through various directions, are the pleasantest I ever saw, little inferior to Leasowes; and, by the beautiful prospects, far superior to Stow gardens. Afterwards we took a view of Lord Clifford's woods, at King's Weston. They are amazingly beautiful: I have seen nothing to equal them in the west of England, and a very few in any other parts.

***Journal*, 7 Jul 1788:** [*Near Doncaster*] After dinner we walked over Mr H's domain, the like to which I never saw in so small a compass. It contains a rabbit-warren, deer, swans, pheasants in abundance, besides a fish pond and an elegant garden. Variety indeed!

***Journal*, 24 Sep 1788:** I took a walk in Miss Goldney's garden at Clifton. Nothing can be more pleasant. But what is most remarkable is, the long terrace walk, commanding a most beautiful prospect, and the grotto, the largest and most beautiful in its kind that I ever saw. It is admirably laid out, and decorated with a surprising variety of shells and glittering fossils; the procuring and placing of which (we were informed) took the late Mr Goldney above twenty years. And he has left it all!

Wales

***Journal*, 24 Aug 1769:** [*Near Caerphilly / Trevecka*] In the evening several of us retired into the neighbouring wood, which is exceeding pleasantly laid out in walks, one of which leads to a little mount, raised in the midst of a meadow, that commands a delightful prospect. This is Howell Harris's work, who has likewise enlarged and beautified his house; so that with the gardens, orchards, walks, and pieces of water that surround it, it is a kind of little paradise.

***Journal*, 25 Aug 1769:** We rode through a lovely country to Chepstow. I had designed to go straight on, but yielded to the importunity of our friends to stay and preach in the evening. Meantime, I took a walk through Mr Morris's woods. There is scarce any thing like them in the kingdom. They stand on top, and down the side, of a steep mountain, hanging in a semicircular form over the river. Through these woods abundance of serpentine walks are cut, wherein many seats and alcoves are placed; most of which command a surprising prospect of rocks and fields on the other side of the river.

***Journal*, 10 Aug 1784:** [*Near Monmouth*] I took a walk to what is called the Bowling-Green House, not a mile from the town. I have hardly seen such a place before. A gravel-walk leads through the most beautiful meadows, surrounded on all sides by fruitful hills, to a gently rising ground, on the top of which is a smooth Green, on which the Gentry of the town frequently spend the evening in dancing. From hence spread various walks, bordered with flowers; one of which leads down to the river, on the back of which runs another walk whose artless shades are not penetrated by the sun. These are full as beautiful in their kind, as even the hanging-woods in Brecknock.

Scotland

***Journal*, 7 May 1761:** I rode over to Sir A. Grant's, near Monymusk, about twenty miles north-west from Aberdeen. It lies in a fruitful and pleasant valley, much of which is owing to Sir Archibald's improvements, who has ploughed up an abundance of waste ground, and planted some million trees. His stately old house is surrounded by gardens, and rows of trees, with a clear river on one side. And about a mile from his house he has laid out a small valley into walks and gardens, on one side of which the river runs. On each side rises a steep mountain; one rocky and bare, the other covered with trees, row above row, to the very top.

***Journal*, 18 May 1770:** [*Near Dunbar*] We rode over to the Earl of Huntingdon's seat, finely situated between two woods. The house is exceeding large and pleasant, commanding a wide prospect both ways; and the Earl is cutting walks through the woods, and much enlarging and beautifying his garden.

***Journal*, 8 Jun 1782:** I walked with a friend to Mr Lesley's seat, less than a mile from the city [*of Aberdeen*]. It is one of the pleasantest places of the kind I ever saw, either in Britain or Ireland. He has laid his gardens out on the side of a hill, which gives a fine prospect both of sea and land: and the variety is beyond what could be expected within so small a compass.

Ireland

***Journal*, 5 Jul 1756:** In the afternoon we came to Hollymount [*near Galway*], some years since one of the pleasantest places in

Ireland. Dr Vesey, the then archbishop of Tuam, fixed on this spot, nine miles from his See, and built a neat commodious house on a little eminence, laid out fruit and flower gardens round it, brought a river to run through them, and encompassed the whole with walks and groves of stately trees. When he had finished his plan, round a stone pillar, which stands in a bason surrounded by a green plat of ground, he placed the following inscription [*translated from the original Greek*]:

'You must leave behind these grounds, this house, and your charming wife, with your numerous and handsome progeny, the endearing offspring of their beloved mother! And all of those trees which you are planting, none, except the loathed cypress, will follow you, their short-lived owner, to the tomb.'

***Journal*, 27-9 May 1773:** I went on to Londonderry. I was invited to see the Bishop's palace, (a grand and beautiful structure,) and his garden, newly laid, and exceeding pleasant. I innocently gave some offence to the gardener, by mentioning the English of a Greek word. But he set us right, warmly assuring us that the English name of the flower is not Crane's bill, but Geranium!

***Journal*, 5 Jun 1773:** [*At Armagh*] I walked over the fine improvements which the Primate has made near his lodge. The ground is hardly two miles round; but it is laid out to the best advantage. Part is garden, part meadow, part planted with shrubs or trees of various kinds. The house is built of fine white stone, and is fit for a nobleman. He intends to carry away a bog which lies behind it, and have a large piece of water in its place. He intends also to improve the town greatly, and to execute many other grand designs; I doubt too many even for a primate of Ireland, that is above seventy years old!

***Journal*, 19 May 1775:** About one I preached at Ballinrobe [*near Galway*] ... In the afternoon I took a view of the castle.

Colonel Cuffe's father took great delight in this place, laid out beautiful gardens, and procured trees of all sorts, from all parts of the kingdom. Part of these placed on the slope of the hill, (at the side of which runs the river,) form a lovely wilderness, at the end whereof are regular rows of elms. But the Colonel has no pleasure therein. So all is now swiftly running to ruin.

***Journal*, 16 Jul 1778:** I went with a few friends to Lord Charlemount's, two or three miles from Dublin. It is one of the pleasantest places I have ever seen: the water, trees, and lawns, are so elegantly intermixed with each other; having a serpentine walk running through a thick wood on one side, and an open prospect both of land and sea on the other. In the thickest part of the wood is the Hermitage, a small room, dark and gloomy enough. The Gothic temple, at the head of a fine piece of water, which is encompassed with stately trees, is delightful indeed. But the most elegant of all the buildings is not finished: the shell of it surprisingly beautiful, and the rooms well contrived both for use and ornament.

***Journal*, 6 May 1783:** I waited on Lady Arabella Denny, at the Black Rock, four miles from Dublin. It is one of the pleasantest spots I ever saw. The garden is everything in miniature. On one side is a grove, with serpentine walks; on the other a little meadow and a green-house, with a study (which she calls her chapel) hanging over the sea. Between these is a broad walk, leading down almost to the edge of the water; along which run two narrow walks, commanding the quay, one above the other.

***Journal*, 3-4 Jul 1787:** A few friends took me to Merino, a seat of Lord Charlemount's, four miles from Dublin. It contains a lovely mixture of wood, water, and lawns on which are several kinds of foreign sheep, with a great plenty of peacocks; but I could not hear any singing birds of any kind. I a little wondered at this, till I afterwards recollected, that I had not heard any singing bird, not even a lark, a thrush or a blackbird, within some miles of Dublin …

I spent an hour at the New-Dargle, a gentleman's seat four or five miles from Dublin. I have not seen so beautiful a place in the kingdom. It equals the Leasowes in Warwickshire; and greatly exceeds them in situation; all the walks lying on the side of a mountain, which commands all Dublin bay, as well as an extensive finely-variegated land-prospect. A little river runs through it, which occasions two cascades, at a small distance from each other. Although many places may exceed this in grandeur, I believe none can exceed it in beauty.

Isle of Man

***Journal*, 30 May-1 Jun 1781:** I embarked on board the packet-boat for the Isle of Man. We had a dead calm for many hours: However, we landed at Douglas on the Friday morning … Before dinner, we took a walk in a garden near the town, wherein any of the inhabitants of it may walk. It is wonderfully pleasant; yet not so pleasant as the gardens of the Nunnery, (so it is called,) which are not far from it. These are delightfully laid out, and yield to few places of the size in England.

Channel Islands

***Journal*, 16 Aug 1787:** [*Guernsey*] I had a very serious congregation at five, in a large room of Mr de Jersey's house. His gardens and orchards are of a vast extent, and wonderfully pleasant; and I know of no Nobleman in Great Britain that has such a variety of the most excellent fruit; which he is every year increasing, either from France, or other parts of the Continent. What quantity of fruit he has, you may conjecture from one sort only: this summer he gathered fifty pounds of strawberries daily, for six weeks together.

***Journal*, 4 Sep 1787:** [*Jersey*] The storm continued, so we could not stir. I took a walk to-day, through what is called the New-Ground, where the gentry are accustomed to walk in the evening. Both the upper ground, which is as level as a bowling green, and the lower, which is planted with rows of trees, are wonderfully beautiful.

Holland

***Journal*, 19 Aug 1786:** We took a walk in Haerlem wood. So delightful a place I scarce saw before. I judged it to be about a mile broad and two or three miles deep. This is divided into almost innumerable walks, some broad and some narrow, but diversified in a wonderful manner, and skirted with elegant houses on both sides.

***Journal*, 25 Aug, 1786:** In the afternoon we took a view of a widow lady's gardens in the suburbs of Utrecht. I believe, from the house to the end of the grand vista is about a mile. I think the gardens are not half as broad; but such exquisite beauty and symmetry I never saw before. In grandeur it is not to be named with a few places in England; but in elegance and variety, I verily believe it equals, if not exceeds, any place of the size in Europe.

Wesley preaching his last sermon

Acknowledgements

Very special thanks are offered to Ralph Waller for his encouragement and help throughout the gestation of this book, and for generously providing the first chapter on the background to John Wesley's life.

Grateful thanks are also due to Sylvia Oliver for valuable help with the preparation of chapters 2–6, and to Adrian Bullock for kindly agreeing to take on the detailed technical stages leading to the final publication, and carrying them out so effectively.

Bibliography and references

If you have found this book interesting and would like to read a little more about John Wesley and his life in the eighteenth century, here are some recent books that can be readily obtained:

Bewes, Richard (comp), *Wesley country*, Bible Matters with Creative Publishing, 2003

Cheetham, Keith, J., *On the trail of John Wesley*, Luath, 2003

Hattersley, Roy, *John Wesley, A brand from the burning: the Life of John Wesley*, Abacus, 2005

Madden, Deborah (ed), *Inward and outward health: John Wesley's holistic concept of medical science, the environment and holy spirit*, Epworth Press, 2008

Wallace, Charles Jr. (ed), *Susanna Wesley: the complete writings*, Oxford University Press, 1997

Waller, Ralph, *John Wesley, a personal portrait*, SPCK, 2003

The following books can be obtained through inter-library loans, or as inexpensive second-hand copies:

Baker, Frank (ed), *John Wesley: letters*, Oxford University Press, 1982

Clarke, Adam, *Memoirs of the Wesley family*, London, 1836

Gill, Frederick, *In the steps of John Wesley*, Lutterworth, 1963

Green, V. H. H., *Young Mr Wesley*, London, 1963

Pearce, John, *The Wesleys in Cornwall*, Bradford Barton, 1964

Pudney, John, *John Wesley and his world*, Thames & Hudson, 1978

Tyerman, L., *The life and times of the Reverend John Wesley*, London, 1890

Wesley, John, *A plain account of Christian perfection*, Epworth Press, 1970

Whiteley, J. H., *Wesley's England*, Epworth Press, 1945
Williams, A. H. (ed), *John Wesley in Wales*, University of Wales, 1971

In addition, various complete and abridged modern or older editions of Wesley's *Journal* are also readily available.

As well as the previously suggested reference book for the climate of the eighteenth century (John Kington's *Climate and Weather*), interesting comments on the weather and natural history from the late 1760s to early 1790s can be found in *The Natural History of Selborne* by Gilbert White which is readily available in several editions.

Quotations

Quotations appearing in this book have been taken from the following sources:

Journal quotations

The Works of John Wesley, Methodist Book Room, London, 1872, reprinted by Baker Books, Michigan, 2007

Letter quotations

The Letters of John Wesley, A.M., edited by J. Telford, Epworth Press, London, 1931, reprinted in 1960

Dates of quotations from the *Journal* and Letters are given in full to avoid confusion, and to facilitate their location elsewhere.

Illustrations

The illustrations in this book have been taken from *Wesley his own Biographer*, published by C. H. Kelly, London, 1891

Index